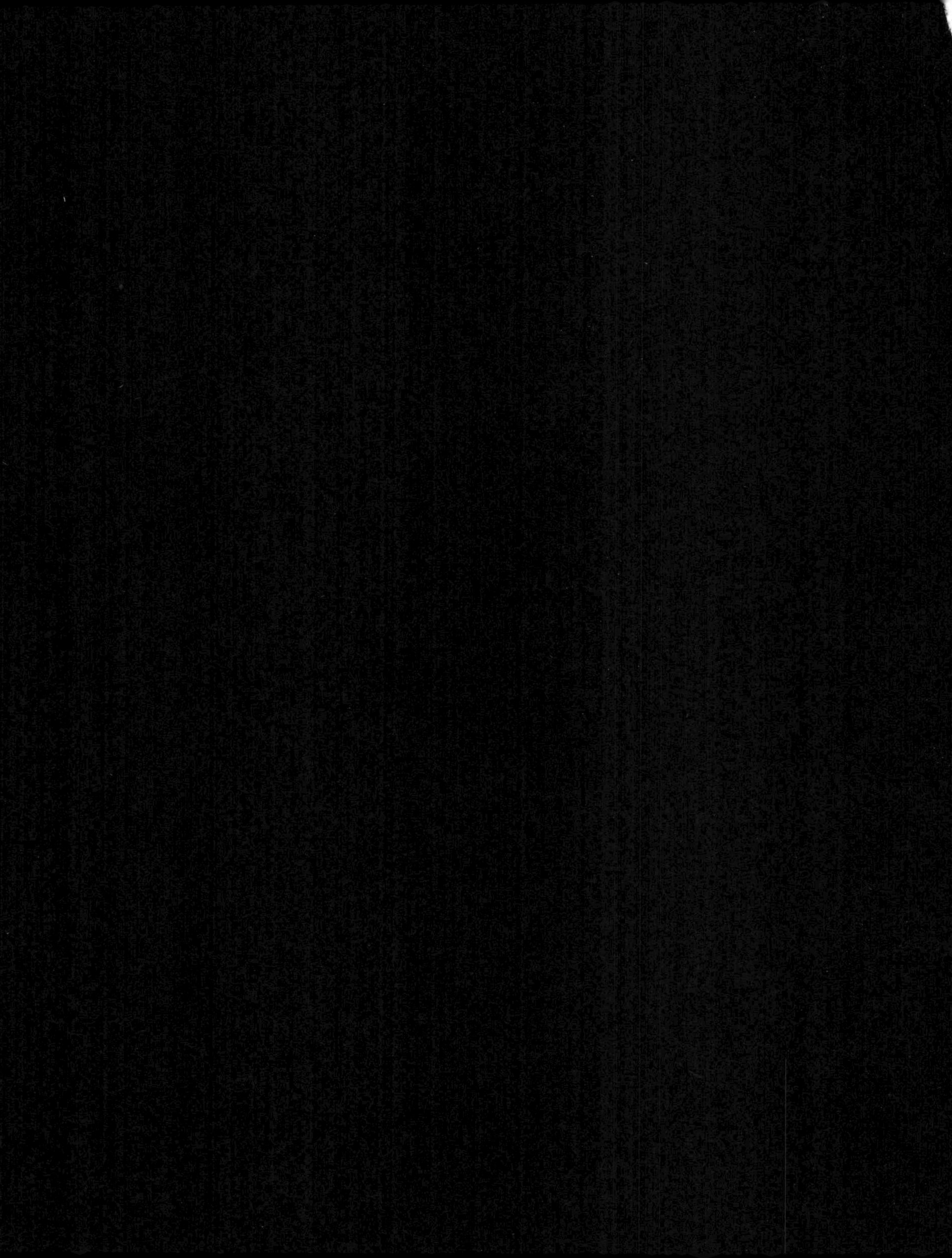

GERMAN AIRCRAFT
OF WORLD WAR I
1914–18

GERMAN AIRCRAFT OF WORLD WAR I
1914–18

EDWARD WARD
& RONNY BAR

amber
BOOKS

First published in 2022

Published by Amber Books Ltd
United House
London N7 9DP
United Kingdom
www.amberbooks.co.uk
Facebook: amberbooks
Instagram: amberbooksltd
Twitter: @amberbooks
Pinterest: amberbooksltd

ISBN: 978-1-83886-112-4

Editor: Michael Spilling
Designer: Mark Batley
Picture research: Terry Forshaw

Printed in Italy

Contents

Introduction

During World War I Germany's air forces transformed from a somewhat ineffectual collection of aircraft and airships into a fiercely effective air arm. By the end of the war Germany led the rest of the world in virtually every branch of military aviation.

The most lasting legacy of Germany's air operations in World War I remains the fighter aircraft. In the Fokker Eindecker, Germany possessed the world's first real fighter and its effect was immediate and decisive. The idea of dedicated fighter squadrons was formulated in early 1916 as the Eindeckers, assigned in ones and twos to general purpose Feldfliegerabteilungen units, detached to form exclusively fighter-equipped squadrons that would soon become known as the formidable Jagstaffeln. The greatest exponent of the fighter aircraft during this formative period was Oswald Boelcke and his eight points for tactical air fighting, the 'Dicta Boelcke', was distributed throughout the Luftstreitkräfte and taught at the fighter school, which he had also suggested founding. The success of German fighter pilots was at its greatest when their equipment matched their

The monoplane Junkers D.I was entering service as the war ended. With its advanced cantilever wing and all-metal construction the D.I represented a harbinger of future military aircraft design, though the features that it pioneered would not become commonplace for 20 years.

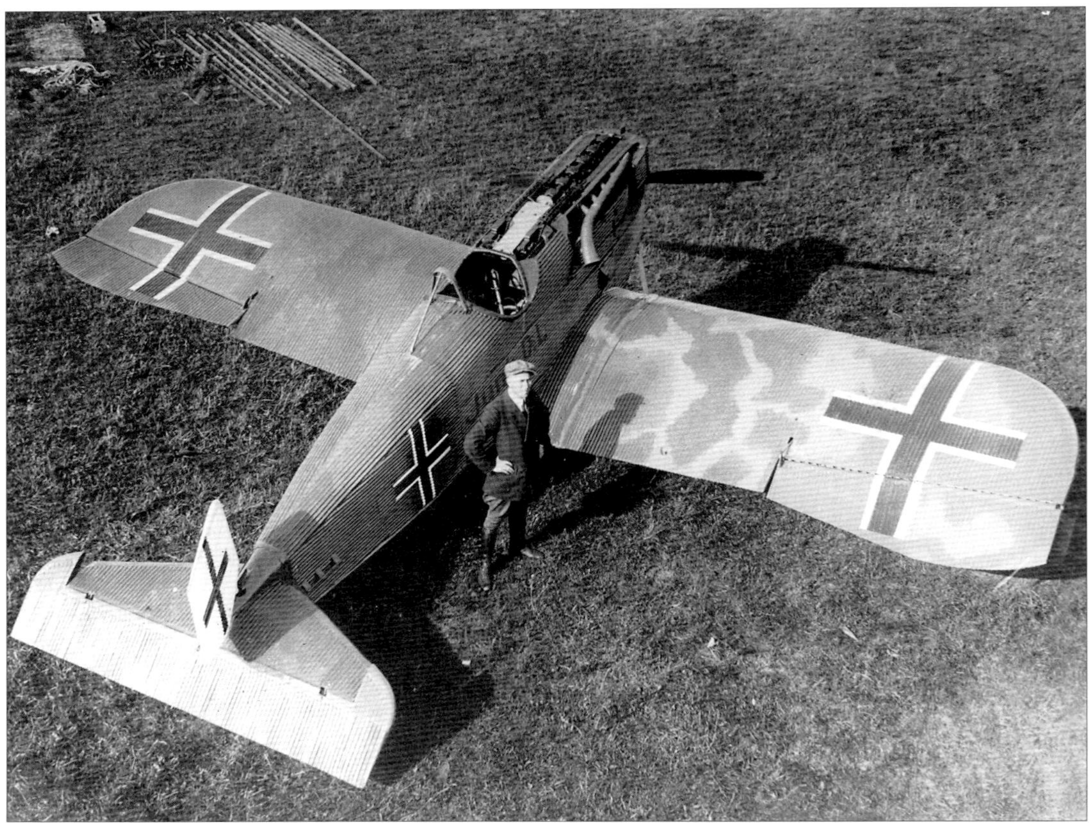

One of the earliest large 'battle planes' of the German Army, the AEG G.II was slow and cumbersome and only around 20 were built. It is chiefly remembered today as the first aircraft flown operationally by Manfred von Richthofen, soon to gain fame as the 'Red Baron'.

skill, thus the Eindecker led to the 'Fokker Scourge', the preeminence of the Albatros fighters in early 1917 resulted in the RFC's 'Bloody April' and the sublime performance of the Fokker D.VII in 1918 led to the Allies', demand that all examples be handed over to them when the war came to a close.

During the last two years of war, the Jagdstaffeln had evolved into the so-called 'Flying Circuses', named in part because of their highly mobile nature, utilizing tent hangars and frequently on the move. Always outnumbered in the West in terms of total amount of aircraft at the Front, the Jagdstaffeln could be moved quickly to any point where they were required. But the 'circus' quality also applied to their aircraft, never before or since has an air force been so gaudily marked in such a dazzling variety of individual colour schemes. Chief among the Jagstaffeln was Jasta 11, commanded by Manfred von Richthofen, the 'Red Baron', who adopted an all-red aircraft and became the war's most successful fighter pilot.

The tendency towards lavish individual decoration was not solely the preserve of the fighter units, as reconnaissance, ground attack, and even large bombers were all lavishly decorated according to the individual whims of their crew. Most of the Luftstreitkräfte's fighter units were Prussian, but the federal nature of the German nation meant that the individual states such as Bavaria maintained armies that were technically separate entities with their own supply chains and equipment. Over the Eastern Front, aerial activity was at a lower ebb and this was used to Germany's advantage as fighter types that struggled over Flanders were able to operate successfully in the East.

Strategic bombers

Germany also pioneered the use of strategic bombers, being the first to drop a bomb on an enemy capital city and first to pursue long-range bombing with rigid airships (the vast and much-feared Zeppelins), and later with conventional aircraft such as the Gothas and enormous Zeppelin-Staakens. In the tactical arena, the Luftstreitkräfte fielded arguably the world's finest ground attack types, the CL series and the armoured J-types (precursor of the Soviet Union's World War II *Sturmovik*) and operated close-support aircraft in a more effective and co-ordinated manner than the Western Allies. Meanwhile, the activities of German reconnaissance aircraft were carried out largely unhindered due to their exceptional altitude performance, the outstanding Rumpler C.IV proving almost impossible for Allied fighters to intercept.

As one of Europe's major land powers, maritime aircraft were less numerous than their Army counterparts, but the Kaiserliche Marine operated fighter squadrons at the Front as well as superlative floatplane fighters that duelled with British flying boats far out over the North Sea. The Navy were also the first to take one of the 'giant' bomber aircraft into action and were the primary operators of Zeppelins throughout the war.

SINGLE-SEAT FIGHTERS

Fokker Eindeckers dominated aerial combat when the first fighters appeared in 1915, though the numbers fielded would be insignificant when compared to the vast production quantities of the mid-war Albatros fighters and the superlative Fokker D.VII of 1918. A large variety of other less commonplace fighters of varying quality were also operated by the Luftstreitkräfte.

This chapter includes the following aircraft:

- Albatros D.I & D.II
- Albatros D.III
- Albatros D.V & D.Va
- Fokker early D-types
- Fokker D.III
- Fokker D.V
- Fokker D.VI
- Fokker D.VII
- Fokker D.VIII
- Halberstadt D.II, D.III & D.V
- Junkers D.I
- Roland D.II & D.IIa
- Roland D.VIa & D.VIb
- Pfalz D.III & D.IIIa
- Pfalz D.VIII
- Pfalz D.XII
- Siemens-Schuckert D.I
- Siemens-Schuckert D.III
- Siemens-Schuckert D.IV
- Fokker Dr.I
- Fokker Eindecker
- Pfalz E-types

Future commander of the *Luftwaffe*, Hermann Goering, was serving with Jagdstaffel 27 when he was photographed in a Fokker Dr.I in early 1918. Goering ended the war with 22 official victories.

Albatros D.I & D.II

Although produced in comparatively small numbers, on its debut the D.I was arguably the world's finest fighter aircraft and heralded the vast swarms of Albatros fighters that would dominate the Jastas during 1917.

Designed by Robert Thelen as a replacement for the successful Fokker Eindecker, the Albatros D.I was an advanced design that set a precedent by favouring speed and climb performance over outright agility. It was also significant for effectively doubling the installed armament of German fighters with its twin MG 08/15 machine guns synchronised to fire through the nose, outgunning all contemporary Allied fighters. Although two (occasionally three) machine guns had been employed on the earlier Fokker E.IV, the D.I was the first to combine this level of firepower with a genuinely effective airframe and a powerful and reliable engine.

Unconventional design

The D.I was viewed as quite unorthodox on its appearance in the summer of 1916. Its semi-monocoque plywood fuselage was lighter and stronger than the conventional fabric-skinned structure of the time, and could more readily be constructed in an aerodynamic shape, contrasting sharply with that of the Eindeckers that preceded it. The engines selected for the new fighter were the most powerful then available in Germany, the 120kW (160hp) Mercedes D.III or 110kW (147hp) Benz Bz.III, neither of which had seriously been considered for fighter applications before due to their large size. Cooling was achieved by two Windhoff 'ear' radiators mounted on the fuselage sides, spoiling the otherwise smooth streamlined shape.

Flown for the first time in June 1916, the D.I swiftly passed its official tests and began to reach operational units in August, although initial deliveries were painfully slow and it was the autumn before the aircraft was available in any numbers, around 50 having been delivered by November

Albatros D.I
Weight: (gross) 898kg (1976lb)
Dimensions: Length 7.4m (24ft 3in) Wingspan 8.5m (27ft 11in) Heigh: 2.95m (9ft 8in)
Powerplant: One 110kW (150hp) Benz Bz.III 6-cylinder water-cooled in-line piston engine or one 120kW (160hp) Mercedes D.III 6-cylinder water-cooled in-line piston engine
Speed: 175km/h (109mph)
Endurance: 1 hour 30 minutes
Ceiling: 5000m (16,400ft)
Crew: 1
Armament: Two 7.92mm (0.312in) LMG 08/15 'Spandau' machine guns

(by which time production had switched to the D.II).

The first few Albatros D.Is began to arrive at the Front at roughly the same time as the reorganisation of aviation units was taking place, resulting in the creation of the first dedicated German fighter units, the Jagdstaffeln or 'Jastas'. Jasta 2 was

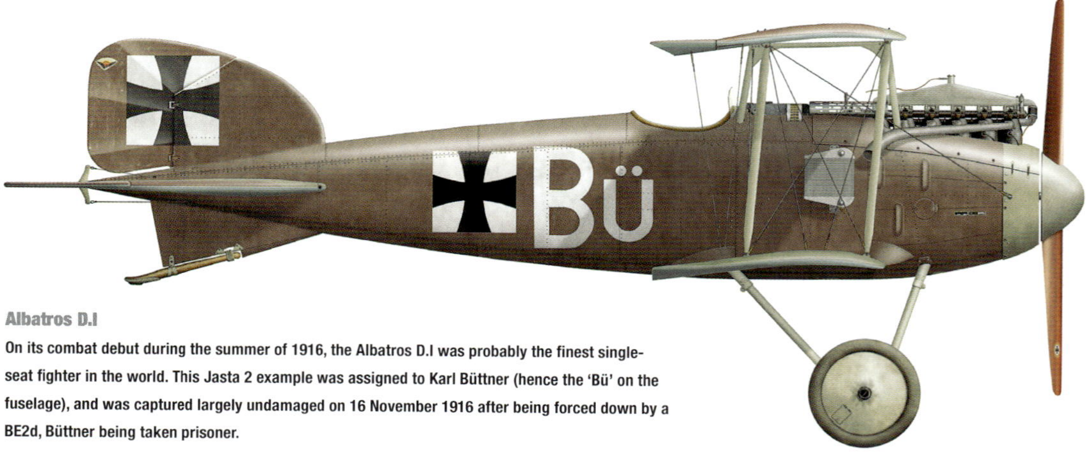

Albatros D.I

On its combat debut during the summer of 1916, the Albatros D.I was probably the finest single-seat fighter in the world. This Jasta 2 example was assigned to Karl Büttner (hence the 'Bü' on the fuselage), and was captured largely undamaged on 16 November 1916 after being forced down by a BE2d, Büttner being taken prisoner.

Albatros D.I

This Albatros D.I is believed to have been on the strength of Jasta 5 and flown by Karl Spitzhoff, the three-pointed star possibly being his personal emblem. At some point the fin and rudder were painted the same blue as the underside, although the white serial number remained on its own little strip of camouflage.

the most successful of the first six Jastas and was initially commanded by the remarkable Oswald Boelcke. At this early stage of their existence Jastas tended to operate a mixture of aircraft types and Jasta 2's initial equipment consisted of one Albatros and two Fokker D.IIIs. New aircraft were generally allocated to the most experienced pilots and Jasta 2 was no exception. Boelcke flew the unit's solitary Albatros D.I and the combination of Germany's most experienced fighter pilot and the finest fighter available was immediately effective, Boelcke scoring seven victories before anyone else had managed a single kill.

The D.I proved as effective as had been expected, with performance comfortably in excess of the Nieuports and DH.2s operated by the Allies at the time. Although both aircraft could outmanoeuvre the Albatros with relative ease, the D.I pilot could dictate the terms on which any combat would be fought and possessed the speed and dive capability to escape at will should it become expedient. Furthermore, with the firepower of two machine guns the D.I could put more bullets into a given target, effectively doubling the possibility of a kill for every accurate burst fired.

Production flaws

Pilots were delighted with the D.I, the only complaint being that the top wing interfered with view from the cockpit to too great a degree, a problem answered by the Albatros D.II. On the D.II the upper wing was lowered 36cm (14in) to a position closer to the pilot's eyeline when looking directly ahead as well as being rigged with greater stagger than the D.I placing it some distance forward relative to the lower wing. In all other regards the two aircraft were identical and the switch to Albatros D.II production was swift.

Later-production D.IIs replaced the fuselage-mounted radiators with a single Teves und Braun unit mounted in the centre section of the upper wing. This was due to IdFlieg banning the use of the 'ear' radiators because, with their bases mounted slightly lower than the engine, if any shot pierced the radiator, gravity was liable to drain the entire cooling system, leading to the engine seizing. The new radiator was aerofoil shaped and matched the profile of the wing, providing a much more aerodynamic cooling system but introduced its own problem in that if holed, scalding water would be blown directly into the pilot's face.

Air superiority

The D.II was the first Albatros fighter to see significant numbers produced, 291 being built (including 16 under licence by Oeffag for Austro-Hungarian use) by the time production switched to the D.III later in 1916. It was by consensus the most formidable combat aircraft available to the Jastas until the D.III started to appear at the Front in December. Operated in concert with the D.I, the performance of both aircraft being basically identical, the Albatros fighters wrested air superiority from the Nieuports and DH.2s of the Allied powers, with many of the greatest German aces scoring a proportion of their early victories on the type.

A notable combat took place on 23 November 1916 when the soon-to-be-famous Manfred von Richthofen engaged in a lengthy dogfight with the RFC's seven-victory ace and Victoria Cross recipient Lanoe Hawker flying a DH.2. Although the DH.2 could out-turn Richthofen in his D.II, eventually low fuel compelled Hawker to break off and run for home, but the Albatros's superior speed prevented Hawker's escape. After firing 900 rounds during the course of the combat, finally the bullets found their mark and Hawker became Richthofen's 11th victim.

Albatros D.II

Weight (Maximum take-off) 888kg (1,954lb)

Dimensions Length: 7.4m (24ft 3in), Wingspan: 8.5m (27ft 11in), Height: 2.64m (8ft 6in)

Powerplant One 120kW (160hp) Mercedes D.III 6-cylinder water-cooled inline piston engine

Speed 175km/h (109mph)

Endurance 1 hour 30 minutes

Ceiling 5200m (17,000ft)

Crew 1

Armament Two 7.92mm (0.312in) MG 08/15 "Spandau" machine guns

Albatros D.II

D.II 386/16 was assigned to Oswald Boelcke, arguably Germany's most influential fighter pilot. In factory finish and devoid of personal markings save for the leader's streamers trailing from the struts, this was the first Albatros airframe to have the radiator mounted in the top wing rather than the fuselage sides. Boelcke lost his life in this aircraft following a collision with another D.II in October 1916.

Albatros D.II

Josef Jacobs was one of the first fighter pilots, flying both Eindeckers and the Fokker D.II before being assigned the considerably more potent Albatros D.II. Emblazoned with his nickname 'Kobes', Jacobs' Albatros hints at the elaborate personal colour schemes soon to be adopted by countless German pilots.

Albatros D.III

The first German fighter to be truly mass-produced, the Albatros D.III was largely responsible for the Jastas' air supremacy during early 1917. Despite its success, the design contained a fatal structural design flaw.

Impressed with the spectacular combat performance of the Nieuport 17, IdFlieg requested that the best aspects of the French aircraft be incorporated into the latest German fighter designs. At Albatros Werke, Robert Thelen decided to mate the Nieuport's sesquiplane layout, in which the lower wing was of a high aspect ratio and featured only one spar, to the fuselage of the D.II. Following closely on the heels of the D.II, the Albatros D.III appeared in the summer of 1916. It is not known when the first flight occurred but it is believed to have taken place in late August or early September. What is known is that the D.III passed its Typenprüfung (type test) on 26 September 1916 and IdFlieg placed an order for 400, the largest German production contract yet issued.

Easily recognisable due to its interplane struts forming a 'vee' shape in place of the parallel struts of the D.I and D.II, the new aircraft was commonly referred to as the 'vee-strutter' by British aircrew. Like the later D.IIs, the radiator was mounted in the top wing, initially in the centre section, but the possibility of showering the pilot with boiling water in the event of combat damage resulted in its sensible relocation to the starboard wing. The D.III boasted better performance than the earlier D.I and D.II, particularly in rate of climb, combining it with a useful improvement in manoeuvrability and the D.III was set to dominate the skies over the Western Front. Unfortunately the D.III hit a deadly structural hurdle almost as soon as it entered service in December 1916.

Fatal design flaw

The major advantage of the sesquiplane layout is the relatively narrow chord of

Albatros D.III

Weight: (gross) 886kg (1949lb)

Dimensions: Length 7.33m (24ft) Wingspan 9.05m (29ft 8in) Height 2.98m (9ft 10in)

Powerplant: One 120kW (160hp) Mercedes D.IIIa 6-cylinder water-cooled inline piston engine

Speed: 175km/h (109mph)

Endurance: 2 hours

Ceiling: 4877m (18,000ft)

Crew: 1

Armament: Two 7.92mm (0.312in) LMG 08/15 'Spandau' machine guns

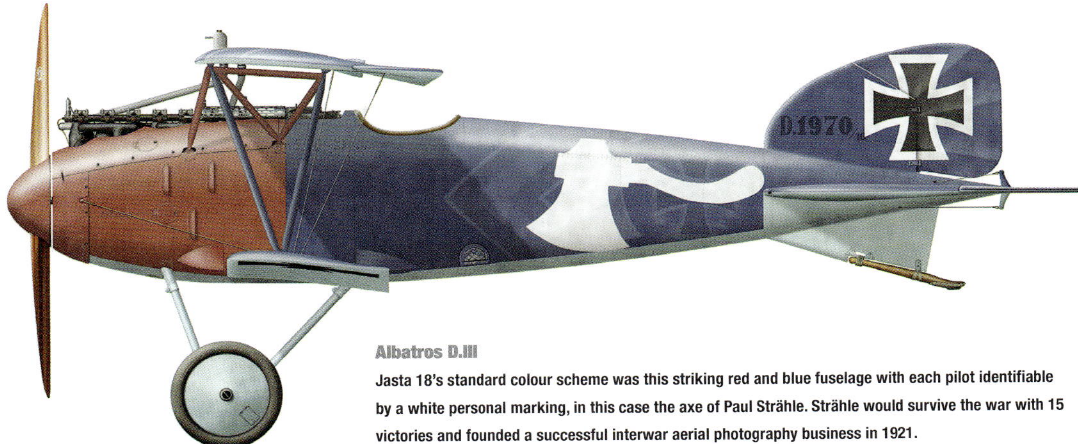

Albatros D.III

Jasta 18's standard colour scheme was this striking red and blue fuselage with each pilot identifiable by a white personal marking, in this case the axe of Paul Strähle. Strähle would survive the war with 15 victories and founded a successful interwar aerial photography business in 1921.

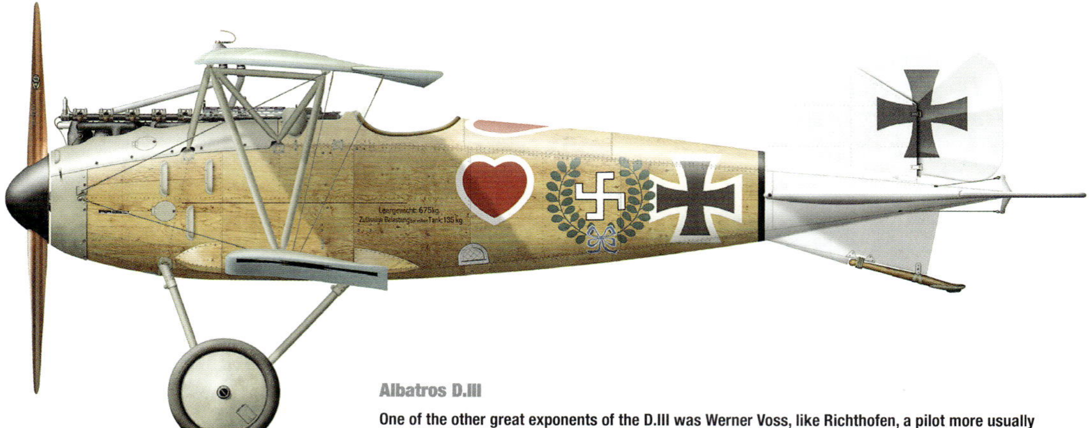

Albatros D.III

One of the other great exponents of the D.III was Werner Voss, like Richthofen, a pilot more usually associated with the Fokker triplane who scored the bulk of his victories in Albatros fighters. Voss flew this D.III emblazoned with swastika and hearts whilst serving with Jasta 2.

the lower wing allows for a much better view downwards, but the Nieuport 17 was known for a problematic tendency under certain flight conditions for the lower wing to twist around its single spar and ultimately fail – a tendency that led to strengthening of the lower wing. However, the Albatros D.III was nearly twice as heavy as the dainty Nieuport and boasted an engine that delivered just under twice the power of the French machine's Le Rhone rotary.

As a result, aerodynamic forces acting on the airframe were significantly higher and the problem was exacerbated on the German aircraft.

On 23 January 1917, a D.III of Jasta 6 suffered a failure of the lower right-hand wing spar. On the following day, the spar of Manfred von Richthofen's brand new D.III cracked and two other pilots were lost due to lower wing failure. Three days later all D.IIIs were grounded pending an investigation

into the problem, forcing the Jastas to revert to the earlier Albatros D.II and Halberstadt D.II.

At the time, the wing spar issue was confusing as the design had been thoroughly load tested before production ensued and found to possess more than adequate strength. Unfortunately, these static tests took place on the ground and the aerodynamic loads that the wing was subject to were not accounted for.

Albatros D.III

The Imperial German Navy contributed five land-based fighter squadrons to the fighting in Europe. This D.III was flown by Josef Rowe of Marine Feld Jagdstaffel I. Rowe transferred to MFJ I in June 1917 from the reconnaissance unit Feld Abteilung 33 and used this D.III to score two victories.

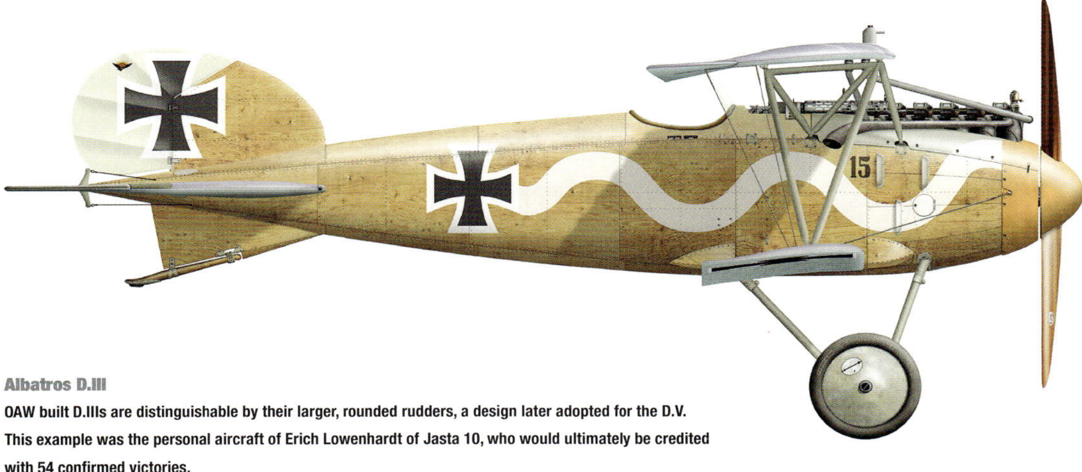

Albatros D.III

OAW built D.IIIs are distinguishable by their larger, rounded rudders, a design later adopted for the D.V. This example was the personal aircraft of Erich Lowenhardt of Jasta 10, who would ultimately be credited with 54 confirmed victories.

Albatros D.III

Despite nearly losing his life to spar failure, Manfred von Richthofen would score 23 of his 80 confirmed victories in the D.III, the most of any individual aircraft type he flew. Richthofen's D.III was nicknamed 'Le Petit Rouge' by Allied flyers.

Albatros D.III

Weight: (gross) 886kg (1949lb)

Dimensions: Length 7.33m (24ft) Wingspan 9.05m (29ft 8in) Height 2.98m (9ft 10in)

Powerplant: One 120kW (160hp) Mercedes D.IIIa 6-cylinder water-cooled inline piston engine

Speed: 175km/h (109mph)

Endurance: 2 hours

Ceiling: 4877m (18,000ft)

Crew: 1

Armament: Two 7.92mm (0.312in) LMG 08/15 'Spandau' machine guns

The wing failures were accordingly suspected to derive from build quality issues at the Albatros' Johannisthal factory. By February, Albatros had introduced a reinforced lower wing that did improve matters somewhat and the grounding order was rescinded although failures continued. For example, on 8 April 1917, the commander of Jasta 4, 20-victory ace and Blue Max winner Wilhelm Frankl, was lost when his D.III suffered a suspected lower wing failure while in combat with Bristol F.2s and broke up in mid-air.

The Red Baron

It is not hard to imagine the effect this structural uncertainty had on front line pilots. Nonetheless, the D.III was extremely successful and was instrumental in allowing the Germans to regain control of the air during early 1917. Pilots were encouraged to 'avoid' prolonged dives in the D.III and the problem, although never eradicated, was largely brought under control. The aircraft was considered generally easy to fly and individual pilots began to rack up impressive scores in the D.III and started to display ever more exuberant personal markings: it was during early 1917 that Manfred von Richthofen first had his D.III painted overall scarlet thus earning his 'Red Baron' soubriquet.

Superior fighter

The D.III was demonstrably superior to all Allied fighters then in service except the Sopwith Triplane and SPAD S.VII, which although outgunned by the Albatros were of roughly similar quality but available only in small numbers in early 1917. During the Battle of Arras in April 1917, German fighters were so dominant that British aircraft losses totalled 245 over the course of the month, with the loss of around 400 aircrew. More than a third of British losses were at the hands of Richthofen's Jasta 11. By contrast, British losses for the entire five months of the Battle of the Somme had totalled 576. The effect on British

morale was reminiscent of the 'Fokker Scourge' a year earlier and the period became known as 'Bloody April'.

Eventually the introduction of the SPAD S.XIII, Sopwith Camel and SE.5a over the next two months or so tipped the balance in favour of the Allies although the D.III would remain in front-line service until the summer of 1918.

Albatros delivered 500 D.IIIs from its Johannisthal factory before production moved to its Ostdeutsche Albatros Werke (OAW) subsidiary at Schneidemühl to allow Johannisthal to concentrate on the D.V. Between June and December 1917 a further 848 D.IIIs would be built by OAW. In Austria-Hungary the D.III was built under licence by Oeffag with a more powerful Austro-Daimler engine.

Oeffag engineers, aware of the wing spar problem, altered the design of the wing to use thicker ribs and spar flanges and resolved the structural problems of the aircraft. They also found that the aircraft was faster if the large propeller spinner was removed. In this form and with other detail changes, approximately 526 D.IIIs were built by the end of the war, proving rugged and popular in

service, 38 serving with the nascent Polish air force during the Polish–Soviet war of 1919–20, where they were thought so highly of that the Poles sent a letter of commendation to the Oeffag factory.

Albatros D.III

Weight: (gross) 886kg (1949lb)
Dimensions: Length 7.33m (24ft) Wingspan 9.05m (29ft 8in) Height 2.98m (9ft 10in)
Powerplant: One 120kW (160hp) Mercedes D.IIIa 6-cylinder water-cooled inline piston engine
Speed: 175km/h (109mph)
Endurance: 2 hours
Ceiling: 4877m (18,000ft)
Crew: 1
Armament: Two 7.92mm (0.312in) LMG 08/15 'Spandau' machine guns

---

Albatros D.III

One of the most flamboyantly decorated military aircraft of all time, this candy-striped OAW-built D.III was flown by Josef Loeser, commander of Jasta 46. Loeser's career was sadly not as spectacular as his aircraft, obtaining two victories before his death in action on 3 June 1918.

Albatros D.V & D.Va

The Albatros D.V's performance was little better than the D.III it replaced, and it suffered from the same major structural problem. Despite this, the D.V and improved D.Va were produced in large numbers and served until the armistice.

Albatros D.V

Friends Fritz Rumey, whose D.V is depicted here, Josef Mai and Otto Könnecke, were three NCO pilots known as the 'Golden Triumvirate' of Jasta 5 due to their prolific victories. Eventually they would shoot down 110 aircraft between them, Rumey being responsible for scoring 45 of these.

During April 1917, the D.III was dominant over the Western Front. Meanwhile, at Albatros Flugzeugwerke, Robert Thelen had designed the D.IV, a thoroughly sensible development of the D.III that returned to the proven wing structure of the D.II and married it to a new, more streamlined elliptical fuselage. Unfortunately the new geared Mercedes engine intended for the D.IV was delayed (and never worked satisfactorily) and by the time it flew, official interest had waned. Meanwhile, IdFlieg, mindful of new Allied fighters in development, sought to maintain the supremacy of their fighters by requesting a 'lightened D.III' from Albatros and the first prototype of the resulting D.V flew later the same month.

Prototype

The prototype D.V was 50kg (110lb) lighter when empty than the D.III and externally similar to its predecessor, the most notable changes being the adoption of the fully elliptical fuselage intended for the D.IV that lacked the flat sides of the D.III, and the upper wing, which was now fitted 12cm (4.7in) closer to the fuselage. The ventral fin was enlarged in area but in all other respects the tailplane was the same as that fitted to OAW-built D.IIIs featuring an elegantly rounded rudder (although the prototype featured the original rudder of early Johannisthal-built D.IIIs). A spinner of greater diameter was fitted to improve the streamlined shape but the wings remained externally indistinguishable from the D.III, the only difference being that the aileron cables were routed through the top wing rather than the lower one. The lower wing also dispensed with the wing-to-fuselage fairing fitted to the D.III. Perplexingly, IdFlieg conducted structural tests on the new fuselage design but not the wings, and in June an order for 900 D.Vs was placed with Albatros.

Albatros D.V

Weight: (gross) 915kg (2017lb)

Dimensions: Length 7.33m (24ft) Wingspan 9.05m (29ft 8in) Height 2.7m (8ft 10in)

Powerplant: One 127kW (170hp) Mercedes D.IIIa 6-cylinder water-cooled inline piston engine

Speed: 170km/h (106mph)

Range: 2 hours

Ceiling: 5700m (18,700ft)

Crew: 1

Armament: Two 7.92mm (0.312in) LMG 08/15 'Spandau' machine guns

D.Vs of the first production batch were fitted with a large headrest but this was usually discarded in service due to the detrimental effect it had on pilot view and it was deleted in later production. Like the D.III before it, the D.V mounted a radiator in the upper starboard wing, although aircraft serving in warmer climates such as Palestine featured two radiators, one either side of the centre section.

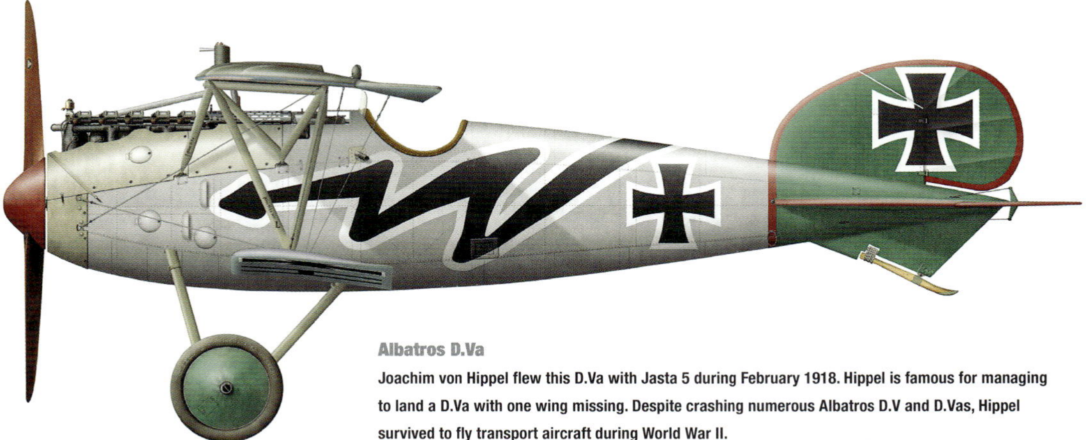

Albatros D.Va

Joachim von Hippel flew this D.Va with Jasta 5 during February 1918. Hippel is famous for managing to land a D.Va with one wing missing. Despite crashing numerous Albatros D.V and D.Vas, Hippel survived to fly transport aircraft during World War II.

Structural failure

The first D.Vs reached the Jastas during June 1917 and proved a disappointment for pilots, initially due to the very modest performance gain delivered by the new aircraft but then followed by the realisation that the structural issues plaguing the D.III were not improved upon by the new Albatros but had, incredibly, worsened. The twisting issue of the single-spar wing remained unchanged but to this was added a propensity for the outboard sections of the upper wing to fail (although this was relatively simply cured with increased bracing) and the lighter fuselage structure was known to crack occasionally on hard landings. More serious still, the D.V featured a weaker structure attaching the lower wing to the fuselage, leading to the entire lower wing tearing away under load, usually when recovering from a dive. IdFlieg immediately conducted belated structural tests of the D.V's wings and,

Albatros D.Va

Jasta 26 adopted this striking striped scheme and this D.Va was flown by Bruno Loerzer during November 1917. In February 1918 Loerzer was promoted to command Jagdgeschwader III, surviving the war with 44 victories.

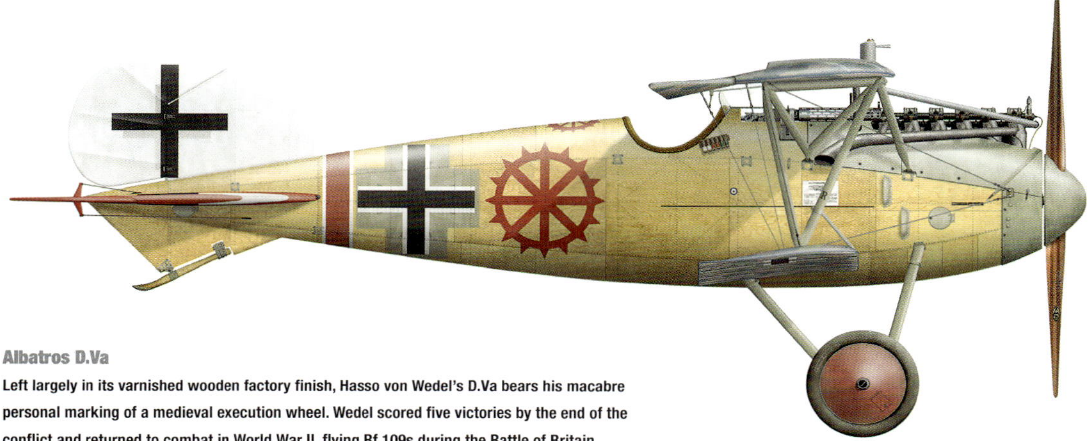

Albatros D.Va

Left largely in its varnished wooden factory finish, Hasso von Wedel's D.Va bears his macabre personal marking of a medieval execution wheel. Wedel scored five victories by the end of the conflict and returned to combat in World War II, flying Bf 109s during the Battle of Britain.

to their great confusion, found the lower wing's strength was more than sufficient. In combat the Albatros was falling further behind the performance of its enemies to such an extent that in July 1917 von Richthofen wrote that the D.V was, "So obsolete and so ridiculously inferior to the English that one can't do anything with this aircraft."

Wing twisting

Albatros introduced the D.Va, which featured a strengthened structure and a return to the D.III's wing design – a process resulting in an aircraft heavier than the D.III, although the use of the high-compression Mercedes D.IIIaü compensated for the weight increase somewhat. The aerodynamic nature of the wing problem was little understood at the time, but Albatros did introduce a small auxiliary strut leading from the forward interplane strut of the 'vee' to the leading edge of the lower wing. This helped prevent the wing from twisting and greatly reduced (although did not entirely eliminate) the propensity for wing failures of the D.Va and it proved successful enough to be retrofitted to D.Vs still in service.

Diving an Albatros remained a cause for concern amongst German pilots, as Ernst Udet noted in his memoirs: "When I was beginning a steep dive I looked at the lower wing and when I noticed it had started to flutter I knew it was time to recover or I would lose my wings." This was hardly an ideal situation for a combat pilot and the failures continued: on 18 February 1918, Hans Joachim von Hippel lost the entire left lower wing of his D.Va at 4000m (13,120ft). Despite this, he was able to retain control of the aircraft and amazingly land the crippled aircraft. Nonetheless, the D.Va continued in production – the Pfalz D.III and D.IIIa could not deliver better performance – and most pilots preferred the Albatros, even with its shaky reputation. With the Fokker Dr.I experiencing its own structural issues, there was no ready alternative to the D.Va, IdFlieg hoping that sheer numbers would make up for the qualitative shortfall of the D.V and D.Va. Nonetheless, many pilots were very successful indeed with the Albatros, reflecting the fact that German fighter pilots were generally more experienced than their Allied counterparts and operating in a target-

rich environment. The appearance of the Fokker D.VII effectively ended the need for the Albatros D.V and production ceased in April 1918, at which point around 1000 D.V and D.Vas were in service on the Western Front. Numbers dwindled over the course of the following months, but the Albatros remained a numerically important fighter for some time. Although most had been reassigned to training duties, a few examples of the final Albatros fighting scout remained operational at the Front until the end of hostilities. Total production numbers for the D.V and D.Va are unknown, but are believed to be around 2500.

Albatros D.Va

Weight: (gross) 937kg (2066lb)

Dimensions: Length 7.33m (24ft) Wingspan 9.05m (29ft 8in) Height 2.7m (8ft 10in)

Powerplant: One 127kW (170hp) Mercedes D.IIIaü 6-cylinder water-cooled inline piston engine

Speed: 172km/h (107mph)

Endurance: 2 hours

Ceiling: 5800m (19,029ft)

Crew: 1

Armament: Two 7.92mm (0.312in) LMG 08/15 'Spandau' machine guns

---

Fokker early D-types

The relatively mediocre performance of Fokker's early biplane fighters was a disappointment after the stunning success of the Eindecker. Nonetheless, they were built in considerable numbers and flown by some of Germany's most renowned aces.

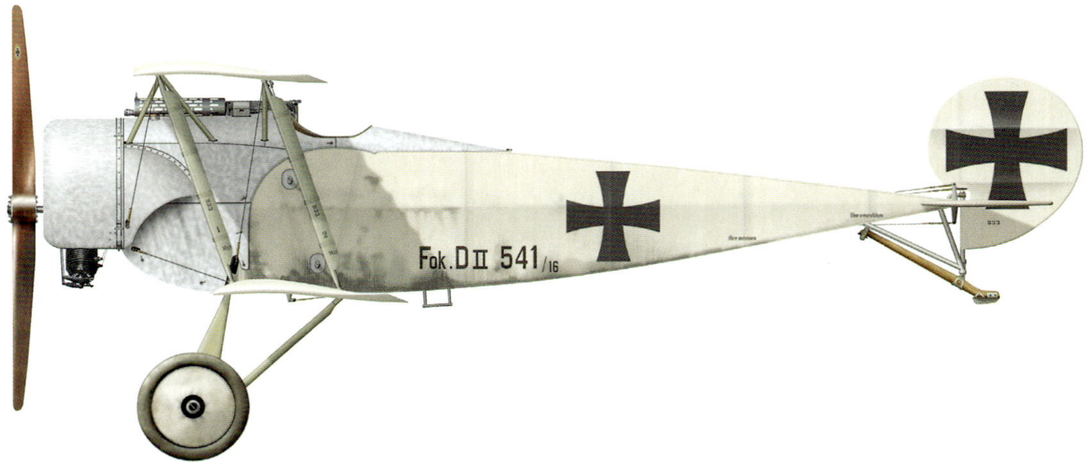

Fokker D.II

Many pilots scored kills with the early Fokker D-types, despite their less than stellar performance. This example was flown by future 48-kill ace Josef Jacobs with Fokkerstaffel-West during 1916. Jacobs later had it painted overall dark blue in an attempt to appear less conspicuous in the air.

Confusingly, the D.I was actually a re-engined version of the D.II – both were developed by designer Martin Kreutzer from the monoplane E.IV, the rear fuselage and tailplane being identical to the earlier machine. The two new biplanes differed in their choice of powerplant, the D.II making use of a 75kW (100hp) Oberursel U.I rotary whereas the D.I featured a water-cooled inline 89kW (120hp) Mercedes D.II. Like the Eindecker, lateral control was achieved by wing-warping rather than ailerons, and unfortunately for pilots at the Front, the performance of the D.I and D.II was little better than the monoplanes they supplanted. Both aircraft arrived during the summer of 1916 and proved to be no match for the superlative Nieuport 17 then dominant over the Front. The Fokker biplanes were rated as inferior to their Albatros and Halberstadt contemporaries and were discarded by front-line Jastas as soon as sufficient quantities of the Albatros D.I and D.II were available. Despite this, the respectable total of 144 and 177 was built of the D.I and D.II respectively, many of which were supplied to Austria-Hungary.

Engine upgrade

In an attempt to improve performance, Fokker fitted a more powerful 120kW (160hp) Mercedes D.III into a strengthened D.I airframe equipped with two machine guns. With wings featuring ailerons for the first time on a Fokker aircraft, the new machine was designated D.IV. The new aircraft also proved inferior to the Albatros fighters and only 40 were delivered, a handful of which were supplied to Sweden.

Fokker D.II

'Dodo' was operational with Kest 4b of the Royal Bavarian Armed forces in late 1916. Note the white-on-black Iron Cross on the rudder and Fokker's distinctive streaky green fabric on the forward fuselage.

Fokker D.II

This D.II was assigned to Fritz Grünzweig of KEK Ensheim during 1916. Grünzweig was amongst the first fighter pilots to elaborately adorn his aircraft, in this case with a monstrous face and three stylized running legs on the wheels.

Fokker D.II

Weight: (gross) 575kg (1268lb)

Dimensions: Length 6.40m (21ft) Wingspan 8.75m (28ft 9in) Height 2.55m (8ft 4in)

Powerplant: One 75kW (100hp) Oberursel U.I air-cooled rotary piston engine

Speed: 150km/h (93mph)

Endurance: 1 hour 30 minutes

Ceiling: 4000m (13,125ft)

Crew: 1

Armament: One 7.92mm (0.312in) LMG 08/15 'Spandau' machine gun

Fokker D.III

Although somewhat more successful than the earlier Fokker D-types, the D.III still possessed indifferent combat performance and its front-line use was relatively brief.

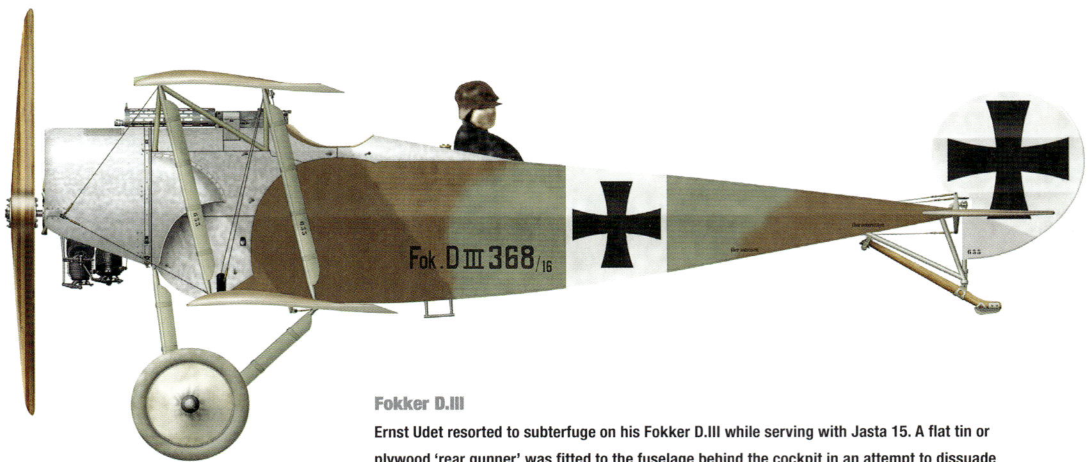

Fokker D.III

Ernst Udet resorted to subterfuge on his Fokker D.III while serving with Jasta 15. A flat tin or plywood 'rear gunner' was fitted to the fuselage behind the cockpit in an attempt to dissuade attempts by Allied fighters to attack from behind.

Mating the strong two-bay wing cellule of the D.I to the fuselage of the D.II, the Fokker D.III utilised Oberursel's two-row 14-cylinder rotary U.III engine which offered 120kW (160hp) and allowed the armament to be increased to two machine guns, matching that of the Albatros D.I. The fuselage structure was strengthened to handle the extra weights and power and after testing in July 1916 IdFlieg placed an order for 50 Fokker D.IIIs, followed by orders for a further 60 in August and 100 in November. Production deliveries began in September, with the first two being ferried to Jasta 2 on the first day of that month.

Poor performer

Service use proved the D.III was an underwhelming aircraft. Oswald Boelcke, leader of Jasta 2 and extremely influential on German fighter development, recommended the D.III be withdrawn to quieter areas of the Front. Nonetheless, the redoubtable Boelcke obtained seven victories in one of the first D.IIIs during the first two weeks of September 1916. Roll control was still affected by wing-warping and as a result manoeuvrability was not particularly impressive. Meanwhile the U.III was an unreliable engine, as had been found when fitted to the earlier E.IV. Prone to overheating and seldom producing its rated power, the U.III could not provide the D.III with sufficient speed.

Furthermore, the D.III was the first aircraft to display Fokker's lacklustre attention to quality control. A D.III fuselage tested to destruction in November 1916 did not meet official specifications and although IdFlieg reprimanded Fokker, his firm's designs would regularly be grounded due to substandard construction. The D.III was forbidden by the Kogenluft for use on front-line duties, although production and usage in secondary units was allowed to continue. Later aircraft were fitted with ailerons and the D.III served with home defence units until late 1917, although 10 D.IIIs supplied to the Netherlands remained in service until 1921.

Fokker D.III

Weight: (gross) 605kg (1334lb)
Dimensions: Length 5.86m (19ft 3in) Wingspan 78.3m (27ft 4in) Height 2.6m (8ft 6in)
Powerplant: One 82kW (110hp) Oberursel Ur.II 9-cylinder air-cooled rotary piston engine
Endurance: 1 hour 30 minutes
Ceiling: 6000m (19,685ft)
Crew: 1
Armament: Two 7.92mm (0.312in) LMG 08/15 'Spandau' machine guns

Fokker D.V

The D.V was not a particularly potent fighter but enjoyed a long and successful second line career as a trainer.

Fokker D.V

The last of Fokker's mediocre mid-war biplane scouts, the D.V was a pleasant enough aircraft to fly but looked somewhat outdated when compared to the latest Albatros fighters. This example, covered with Fokker's trademark streaky fabric over all upper surfaces, was serving with an unidentified Naval unit during 1917.

A development of the D.II, the D.V featured an aerodynamically cleaned-up fuselage with a new cowling completely enclosing the Oberursel U.I engine and a large, blunt spinner over the airscrew hub. The wings were markedly staggered and the top wing was swept back to improve pilot view. Roll control was now achieved by ailerons for the first time on a Fokker fighter, rather than utilising wing-warping, and handling was improved as a direct result. Armament remained a single 'Spandau' machine gun due to the comparatively low power of the engine and the D.V was deemed to offer sufficient improvement over the earlier Fokker biplanes to be ordered into production by IdFlieg in October 1916. In total 216 would ultimately be built, making the D.V the most produced of Fokker's biplane fighters thus far.

Making its combat debut over the Western Front in January 1917, it was clear that the D.V, although offering pleasant handling characteristics, was too modest in performance, particularly climb rate, when compared to its contemporaries. The Jastas much preferred the considerably more powerful Albatros D.I and D.II, which also boasted twice the firepower of the D.V. After seeing relatively brief combat service at the Front, the D.V was relegated to second-line duties and it served as an interceptor on home defence duties with both Army and Naval units. Furthermore, its docile handling characteristics rendered it effective as a fighter trainer, a role it performed until the end of the war. A few D.Vs subsequently returned to the Front in late 1917 for use as conversion trainers at operational Jastas equipping with the new Fokker Dr.I.

Fokker D.V

Weight: (gross) 566kg (1245lb)

Dimensions: Length 6.05m (19ft 10in) Wingspan 8.75m (29ft 1in) Height 2.3m (7ft 7in)

Powerplant: One 75kW (100hp) Oberursel U.I 9-cylinder air-cooled rotary piston engine

Speed: 170km/h (106mph)

Endurance: 1 hour 30 minutes

Ceiling: 5000m (16,400ft)

Crew: 1

Armament: One 7.92mm (0.312in) LMG 08/15 'Spandau' machine gun

Fokker D.VI

The D.VI was an excellent aircraft with fine flying characteristics, but engine issues combined with the terrific success of the contemporary D.VII resulted in it being built only in small numbers and service use was limited.

Consisting essentially of a scaled down set of D.VII wings attached to a modified Dr.I fuselage, two Fokker V.13 prototypes were built during 1917. It is not known when the aircraft was first flown, but both were entered into the Adlershof fighter trial in late January of 1918. Despite one of the new fighters being taxied into an Albatros D.V in foggy conditions, the V.13 was adjudged the finest rotary-engine fighter at the competition and it was ordered into production as the D.VI.

Engine availability

Deliveries of the D.VI began in April 1918 but unfortunately for the D.VI its preferred engine, the 137kW (183hp) Goebel Goe.III, was at an early stage of development and not available in sufficient quantity. Only 12 D.VIs are recorded as having the Goe.III fitted and most production aircraft were forced to utilise the considerably less powerful Ur.II, itself a licence-built Le Rhône 9J, with a concomitant effect on performance. Furthermore, reliability of German rotary engines was suffering as the war progressed due to the lack of the castor oil required to lubricate them.

A synthetic alternative named 'Voltol' could not match castor oil and engine life plummeted. Ultimately Fokker ceased production of the D.VI in August at the 59th example to concentrate all its efforts on the superlative D.VII, for which there was insatiable demand.

Issued in small numbers to several front-line units, the D.VI was delightful to fly, offering greater manoeuvrability and marginally better performance than the D.VII at low altitude, but by September 1918 all remaining examples had been relegated to training and home-defence units.

Fokker D.VI

Weight: (gross) 585kg (1290lb)
Dimensions: Length 6.25m (20ft 6in) Wingspan 7.66m (25ft 1in) Height 2.55m (8ft 4in)
Powerplant: One 82kW (110 hp) Oberursel Ur.II 9-cylinder air-cooled rotary piston engine
Speed: 197km/h (123mph)
Range: 300km (186 miles)
Ceiling: 6000m (19,685ft)
Crew: 1
Armament: Two 7.92mm (0.312in) LMG 08/15 'Spandau' machine guns

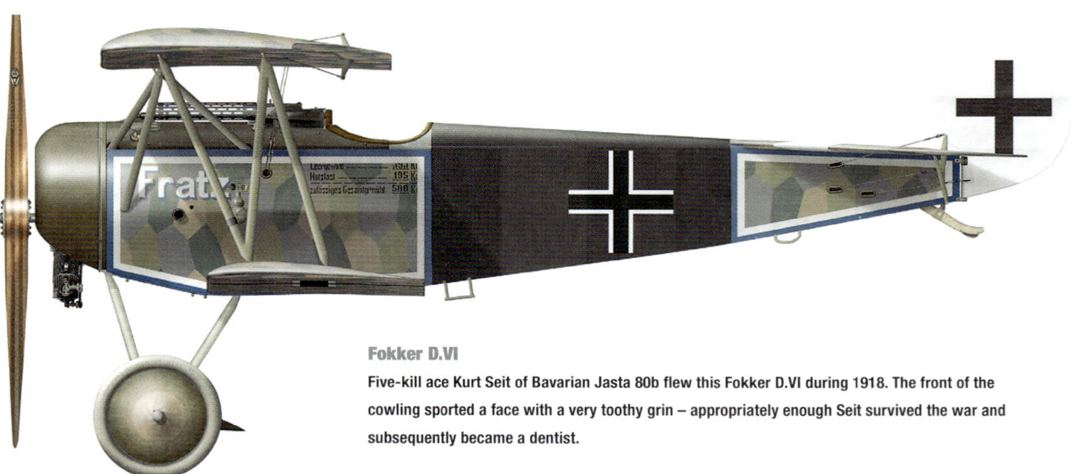

Fokker D.VI

Five-kill ace Kurt Seit of Bavarian Jasta 80b flew this Fokker D.VI during 1918. The front of the cowling sported a face with a very toothy grin – appropriately enough Seit survived the war and subsequently became a dentist.

Fokker D.VII

By consensus the most formidable fighter to see widespread service during the conflict, the Fokker D.VII transformed the fortunes of both the Jagdflieger and the Fokker company.

One of the terms of the 1918 armistice that would bring the Great War to a close was that Germany surrender "1700 fighting and bombing aeroplanes – in the first place, all D7's (sic) and all night bombing aeroplanes." That the Fokker D.VII should be the only German aircraft, indeed the only German weapon, specifically required to be handed over to the Allies gives an indication of the respect it was accorded by its opponents.

First fighter competition

The D.VII was the successful result of work carried out by the Fokker company on a series of aircraft featuring a thick cantilever wing with a smoothly curved leading edge. It was found that this wing offered greater lift and therefore a more docile stall than the thin wings in general use and, somewhat counterintuitively, also conferred less drag than a thin wing. The aircraft also used a welded steel tube fuselage frame that was a departure from general practice at the time. The prototype of what was to become the D.VII was Fokker's V.11, which was essentially an inline Mercedes D.III-powered version of the rotary powered V.9 that would ultimately become the D.VI. The V.11 flew for the first time in December 1917 and was entered in IdFlieg's First Fighter Competition held during January of 1918. This was the first event of its kind in which the opinion of serving fighter pilots was sought and chief amongst them was Manfred von Richthofen, then at the height of his fame. Richthofen flew the V.11 and found it unpleasant to fly, directionally unstable and oversensitive. Fokker immediately lengthened the fuselage and added a triangular fin in front of the rudder. In this form the V.11 found ready acceptance with both von Richthofen and amongst the other

Fokker D.VII (early production)
Weight: (gross) 906kg (1997lb)
Dimensions: Length 6.95m (22ft 10in) Wingspan 8.9m (29ft 2in) Height 2.75m (9ft)
Powerplant: One 120kW (160hp) Mercedes D.lll 6-cylinder water-cooled inline piston engine
Speed: 188km/h (117mph)
Endurance: 1 hour 30 minutes
Ceiling: 6000m (19,685ft)
Crew: 1
Armament: Two 7.92mm (0.312in) LMG 08/15 'Spandau' machine guns

Fokker D.VII

The first D.VIIs were finished in Fokker's distinctive streaky olive canvas on the fuselage, although a change was soon made on the production line to lozenge fabric. This example was flown by 10-victory ace Willi Gabriel. In 1912 at the age of 19, Gabriel had built and flown his own aircraft and his later experiences as an NCO pilot in the class-ridden *Luftskreitkräfte* inspired the 1966 film *The Blue Max*.

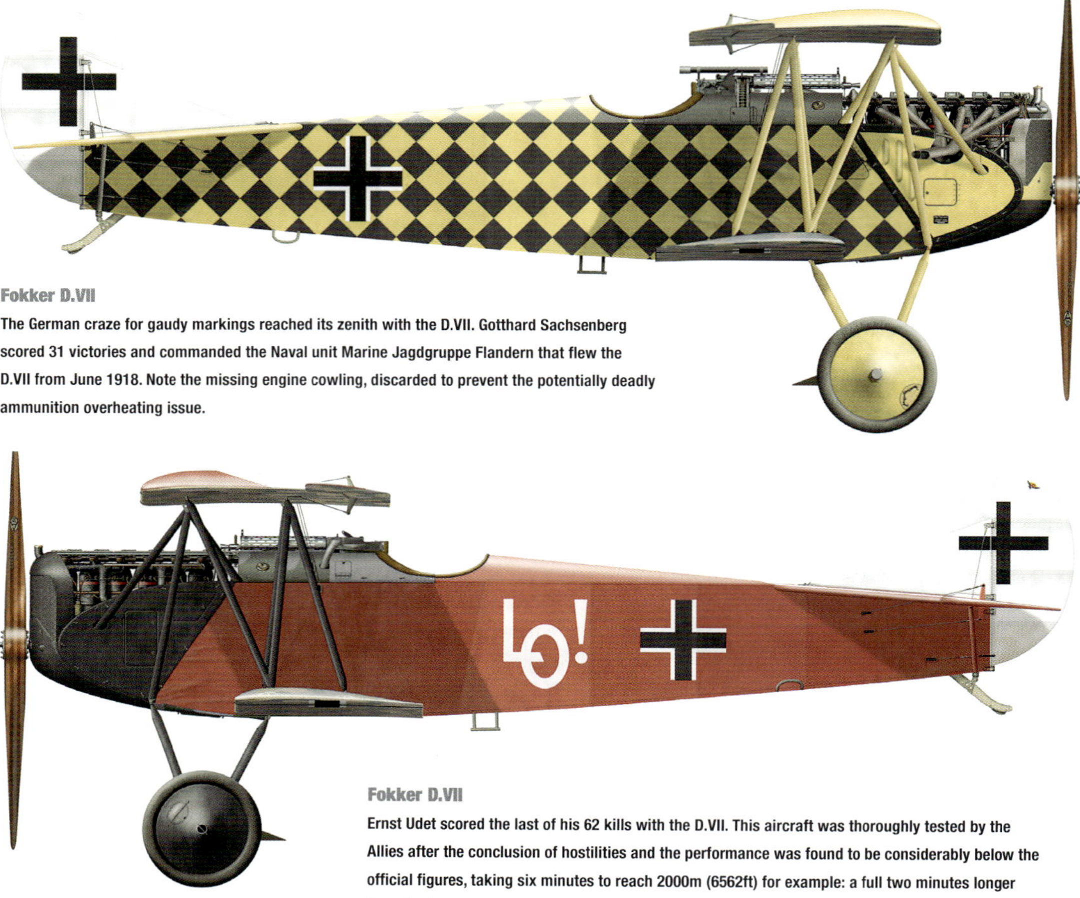

Fokker D.VII

The German craze for gaudy markings reached its zenith with the D.VII. Gotthard Sachsenberg scored 31 victories and commanded the Naval unit Marine Jagdgruppe Flandern that flew the D.VII from June 1918. Note the missing engine cowling, discarded to prevent the potentially deadly ammunition overheating issue.

Fokker D.VII

Ernst Udet scored the last of his 62 kills with the D.VII. This aircraft was thoroughly tested by the Allies after the conclusion of hostilities and the performance was found to be considerably below the official figures, taking six minutes to reach 2000m (6562ft) for example: a full two minutes longer than the factory specification.

pilots, being universally judged the best aircraft of the competition and an order for 400 aircraft designated D.VII was duly placed by IdFlieg.

Fokker was unable to supply such a large order and thus IdFlieg ordered Albatros to build the D.VII under licence, a situation that Anthony Fokker found intolerable as he believed his greatest rival was profiting from his own design, although the five percent royalty Albatros were required to pay Fokker for each D.VII airframe it produced softened the blow somewhat.

Albatros requested working drawings to tool up for production but Fokker had never used detailed drawings in the production process, answering the problem by simply delivering a complete D.VII to the Albatros factory from which proper working drawings were prepared. Both Albatros and its OAW subsidiary produced the aircraft, these differing in detail from each other and from those produced by Fokker, and many parts were not interchangeable between D.VIIs from each production source.

Quality control shortcomings

D.VIIs reached the Front in early May 1918, Jasta 10 being the first unit to receive examples of the new aircraft but Richthofen, who had championed the design at Adlershof, never got the chance to fly it in combat as he died a few days before the first examples reached the Front. Pilots were delighted with the D.VII, as it offered a jump in performance over the Albatros D.Va despite being fitted with the same engine, better manoeuvrability and possessed none of the structural

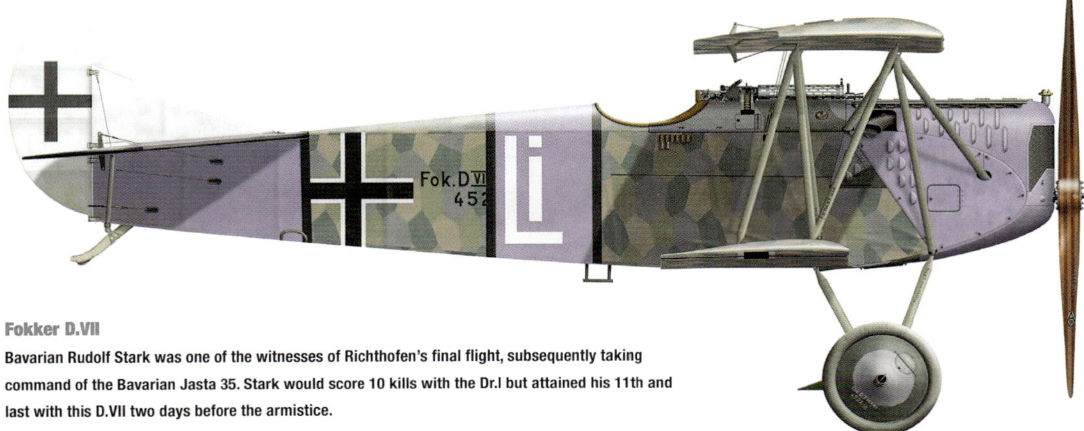

Fokker D.VII

Bavarian Rudolf Stark was one of the witnesses of Richthofen's final flight, subsequently taking command of the Bavarian Jasta 35. Stark would score 10 kills with the Dr.I but attained his 11th and last with this D.VII two days before the armistice.

weakness of the earlier aircraft, although the tedious inevitability of Fokker's slapdash quality control led to some upper wing rib failures and loss of skinning but nothing serious enough to warrant a grounding order as befell the Dr.I. There was never any such problem with D.VIIs built by Albatros or OAW, build quality from both these sources being rated as significantly better than that of Fokker.

The single greatest attribute that the D.VII possessed was its benign handling characteristics courtesy of the thick wing. The D.VII was almost laughably easy to fly, particularly at high altitude and near the stall, which was gentle and easy to recover from, with the aircraft demonstrating a marked reluctance to enter a spin. By contrast, contemporary fighters such as the SPAD XIII and Sopwith Camel stalled sharply and would fall easily into a vicious spin. It was famously remarked that the D.VII made a bad pilot into a good one, and a good pilot into an ace. This was of course an exaggeration, but what the D.VII offered was an aircraft with world-class performance in which the pilot was not constantly concerned with the difficult business of simply flying the aircraft and could concentrate on combat, increasing situational awareness, promoting confidence and encouraging forthright handling. The D.VII was not entirely faultless, however. Early examples placed the ammunition supply too close to the

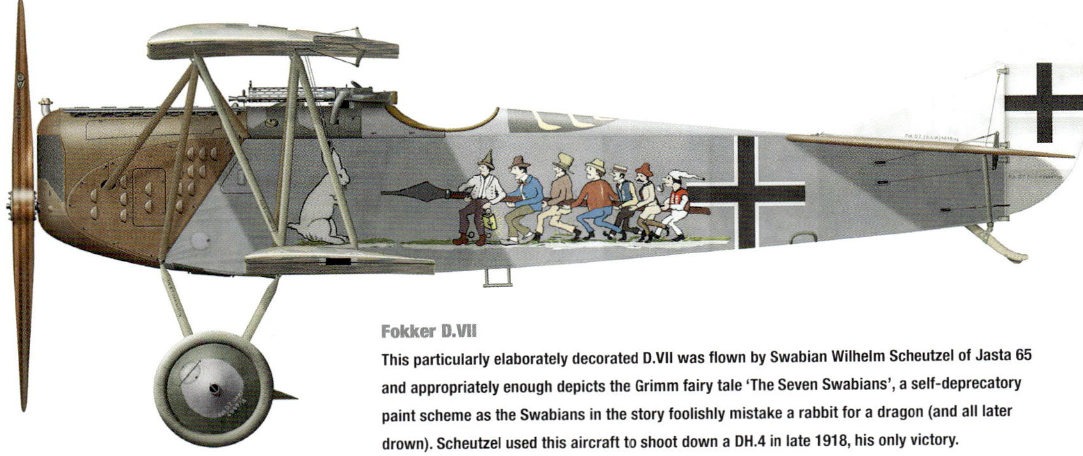

Fokker D.VII

This particularly elaborately decorated D.VII was flown by Swabian Wilhelm Scheutzel of Jasta 65 and appropriately enough depicts the Grimm fairy tale 'The Seven Swabians', a self-deprecatory paint scheme as the Swabians in the story foolishly mistake a rabbit for a dragon (and all later drown). Scheutzel used this aircraft to shoot down a DH.4 in late 1918, his only victory.

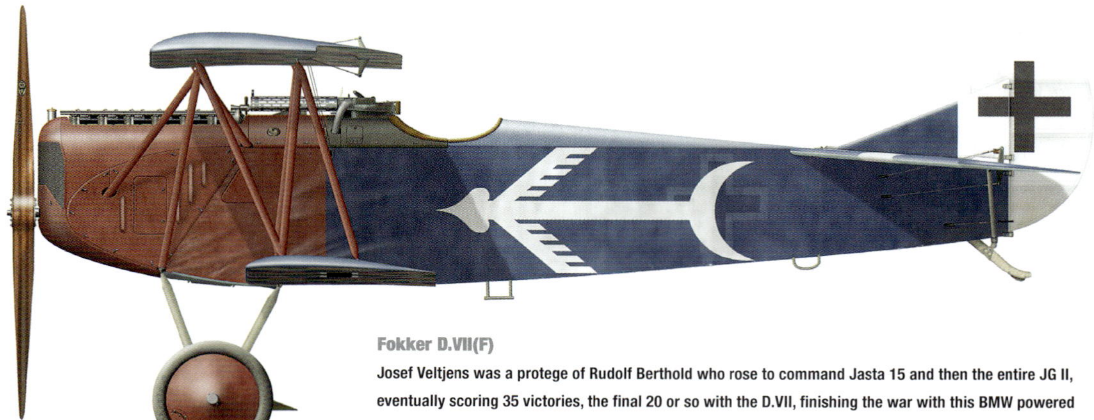

Fokker D.VII(F)

Josef Veltjens was a protege of Rudolf Berthold who rose to command Jasta 15 and then the entire JG II, eventually scoring 35 victories, the final 20 or so with the D.VII, finishing the war with this BMW powered D.VII(F). Veltjens' subsequent career was colourful, becoming an arms dealer between the wars, rejoining the Luftwaffe in 1939 and serving as Goering's personal envoy to both Finland and Mussolini's Italian Social Republic.

engine and there were instances in which bullets 'cooked off' in flight, setting the aircraft alight and leading to the deaths of several pilots including 21-victory ace Fritz Friedrichs. The problem was relatively easily solved by increasing the flow of cooling air through the front fuselage and the use of improved ammunition.

Also problematic was the engine situation. The D.VII made use of the Mercedes D.III and D.IIIaü, although this power plant was getting decidedly long in the tooth by early 1918, and a new engine was desperately needed to match the obvious potential of the airframe. During June just such an engine appeared in the form of the BMW IIIa, the first ever product of the BMW firm. Optimised for high-altitude use, the 'overcompressed' BMW featured a carburettor that adjusted fuel mixture for operation at different heights and a very high compression. Indeed, the compression was high enough that use of full throttle was prohibited below 2000m (6562ft) as it risked premature detonation and damage to the pistons and cylinders. The BMW engine transformed the D.VII's already

impressive performance and conferred upon it a rate of climb superior to any Allied aircraft then in service. Aircraft with the BMW engine were designated D.VII(F), the 'F' standing for Max Fris, who had designed it. Supply issues meant that there were never enough BMW engines to meet demand and D.VIIs with the Mercedes D.III, Mercedes D.IIIaü and BMW III were all produced concurrently right up to the armistice.

Worldwide acceptance

Unsure of what to make of the new boxy aircraft, lacking the sleek lines of the Albatros fighters it replaced, Allied aircrew initially underestimated the capabilities of the D.VII. In combat from May 1918 onwards, the D.VII soon dazzled its opponents, its ability to 'hang on its prop' and remain controllable was remarked upon by many contemporary Allied aviators who had never witnessed anything quite like it before. Production built up very quickly and around 70 front-line units would ultimately be equipped wholly or in part with the D.VII, and over 3000 were built in total, of which 775 were in service in November 1918. Licence

Fokker D.VII(F)

Weight: (gross) 906kg (1997lb)

Dimensions: Length 6.95m (22ft 10in) Wingspan 8.9m (29ft 2in) Height 2.75m (9ft)

Powerplant: One 138kW (185hp) BMW IIIa 6-cylinder water-cooled inline piston engine

Speed: 200km/h (124mph)

Endurance: 1 hour 30 minutes

Ceiling: 7000m (22,900ft)

Crew: 1

Armament: Two 7.92mm (0.312in) LMG 08/15 'Spandau' machine guns

production was also undertaken of an Austro-Daimler powered version by MAG for Austria-Hungary, although only around 50 aircraft were built. After the war ended Anthony Fokker managed to smuggle six trainloads of aircraft and parts out of Germany into his native Netherlands where production of the D.VII continued for eager customers worldwide. Meanwhile D.VIIs handed over in accordance with the terms of the armistice saw squadron service in Belgium and the US. Ultimately the D.VII would operate with the militaries of 19 nations, the last examples in front-line service only being withdrawn by Switzerland in 1938.

Fokker D.VIII

The last Fokker fighter design to see service in World War I, the Fokker D.VIII was a fine aircraft delayed at the very moment it was most sorely needed and saw little action as a result.

Fokker had produced several monoplane fighter prototypes by early 1918 with a single cantilever wing mounted directly to the fuselage. This had invariably led to an unacceptable view for the pilot and so, possibly influenced by the appearance of the French Morane A1 parasol monoplane over the Front, Fokker simply removed the bottom wing of a standard D.VII biplane to see how it would perform with just one wing. The results proved satisfactory and two parasol monoplanes were quickly prepared, the V.26 and V.28, differing from each other only in that the V.28 was strengthened to accept engines of greater power.

Both were present at the Second Fighter Competition during May 1918 and despite the low power of the outdated Oberursel Ur.II fitted to the V.26, its low drag and weight resulted in a more than satisfactory performance, the aircraft being judged only slightly inferior to the Siemens-Schuckert D.III that utilised the complex Sh.III delivering around a third as much power.

This performance created enough of an impression for IdFlieg to place an order for 400 aircraft designated the Fokker E.V. The intention was to power this initial order with either the 108kW (145hp) Oberursel Ur.III or the 119kW (160hp) Goebel Goe.III but neither of these engines was ever available in sufficient quantity and all examples of the Fokker parasol would be powered by the Ur.II.

Fokker E.V

Weight: (gross) 605kg (1334lb)

Dimensions: Length 5.86m (19ft 3in) Wingspan 78.3m (27ft 4in) Height 2.6m (8ft 6in)

Powerplant: One 82kW (110hp) Oberursel Ur.II 9-cylinder air-cooled rotary piston engine

Endurance: 1 hour 30 minutes

Ceiling: 6000m (19,685ft)

Crew: 1

Armament: Two 7.92mm (0.312in) LMG 08/15 'Spandau' machine guns

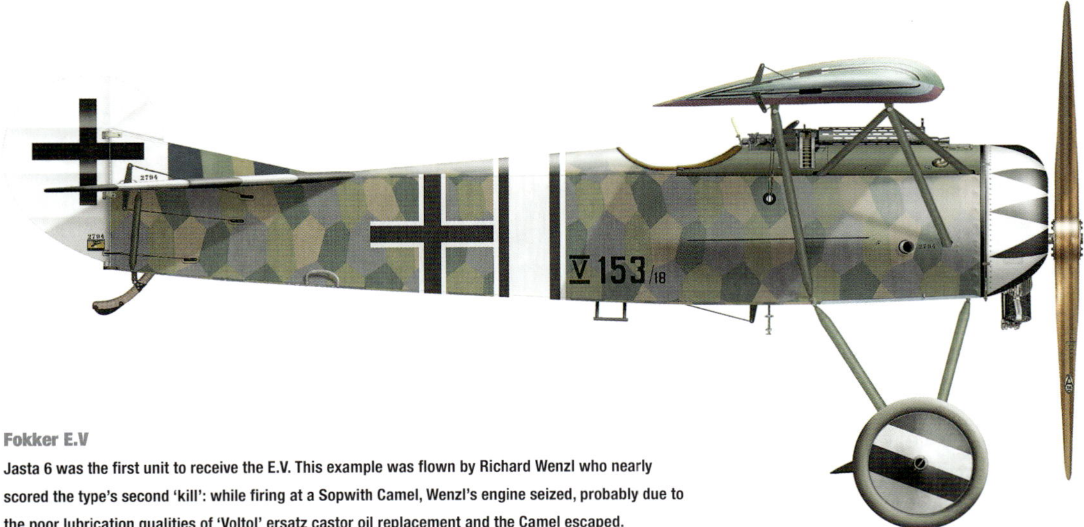

Fokker E.V

Jasta 6 was the first unit to receive the E.V. This example was flown by Richard Wenzl who nearly scored the type's second 'kill': while firing at a Sopwith Camel, Wenzl's engine seized, probably due to the poor lubrication qualities of 'Voltol' ersatz castor oil replacement and the Camel escaped.

Fatalities mount up

The first E.Vs arrived at the Front in late July and on 17 August 1918 Emil Rolff of Jasta 6 scored the type's first victory when he shot down a Sopwith Camel of 203 squadron. Unfortunately, a mere two days later Rolff was killed when the wing of his E.V failed in flight. Two more fatal crashes due to wing failures occurred within a week, the E.V fleet was grounded and IdFlieg launched an investigation. At first the design of the wing was erroneously thought to be at fault, but once again Fokker's cavalier attitude to quality control was to blame – admittedly in this case due to wings built by subcontractor, the Perezina Piano factory. These were found to be sloppily made – some of the wood was green and some affected by dry rot. In a repeat of the Dr.I fiasco, IdFlieg demanded Fokker replace all E.V wings at his own expense and mandated that Fokker staff should directly oversee production at Perezina.

Production resumed in late September, new aircraft being redesignated D.VIII, despite 'D' ostensibly standing for 'Doppeldecker' (biplane), IdFlieg having decided that all fighters would be D-types regardless of how many wings they possessed.

Combat experience

The first D.VIII arrived at the Front on 24 October 1918 but 18 days later the war was over. Despite the brevity of its combat service the E.V/D.VIII had made quite an impression on both sides of the Front. Ernst Udet had demonstrated its qualities by using one to best a Fokker D.VII flown by 25-victory ace Robert von Greim in mock combat, while British Camel ace Edwin Swale was bounced by four E.Vs on 12 August 1918 and stated that they, "Outmanoeuvered him in every way", although he managed to escape.

After the war Poland utilised seven captured examples during the Polish–Soviet War and Stefan Stec used one of these to achieve the first victory of the Polish Air Force when he downed a Ukrainian Nieuport 11 on 29 April 1919.

Fokker D.VIII

Friedrich Altemeier was a talented NCO pilot who scored a total of 21 confirmed victories. His final 'kill' was also the last German air-to-air victory of the war when he shot down an RE8 at 10.50 am on 10 November 1918, a feat he is believed to have attained in this D.VIII.

Halberstadt D.II, D.III & D.V

For a brief period the finest fighter available to the Jagdstaffeln were the Halberstadt single-seaters that supplemented Fokker's biplane fighters, which bridged the gap between the Eindeckers and the arrival of the Albatros Scouts.

Halberstadt D.II

The 810/16 was a D.II built under licence by Hannoversche Waggonfabrik AG and is believed to have served with either Jasta 25 or Halbgeschwader I in Macedonia. The letter 'S' probably refers to the pilot's surname, but who this was remains unknown.

During 1915, Halberstädter Flugzeugwerke developed the world's first fighter powered by a water-cooled inline engine, the D.I. Although production of this aircraft did not go ahead, Halberstadt developed an improved version with radiator moved from the nose to the wing and a built-up 'turtledeck' surrounding a raised pilot's seat to improve the view from the cockpit. In this form the aircraft was redesignated D.II and entered production, the first examples arriving at combat units in early 1916. With its slender fuselage and distinctive strut braced all-moving tailplane, the D.II did not appear particularly sturdy but the aircraft in fact boasted great structural strength, a fact that saw the aircraft returned to service in early 1917 after its front-line career had seemingly ended, when the structural issues of the Albatros D.III first appeared. The Halberstadt also featured ailerons rather than wing-warping, bestowing it with a handling advantage over its Fokker biplane contemporaries. Official figures (if correct) suggest the performance of the D.II was little better than the Eindeckers it replaced, but it was generally acknowledged by pilots to be the best German fighter available until the arrival of the Albatros D.I and it was accorded respect by Allied aircrews. A July 1916 IdFlieg report comparing the D.II to the Fokker E.IV noted that, "The Halberstadt … is well regarded – especially praised are its ability to climb and manoeuvre. It is decidedly preferred to the 160hp Fokker." Such noted aces as Manfred von Richthofen and Oswald Boelcke both scored victories in the D.II and its success belied the fact that the trivial total of 65 D.IIs was built, most of those under licence by Aviatik and Hannoversche Waggonfabrik, which were confusingly designated Aviatik D.I and Hannover D.I respectively.

Meanwhile, Halberstadt had developed the D.III, of which 50 were produced, which was basically the same aircraft but with an Argus As.II engine of the same six-cylinder layout and power output in place of the original Mercedes D.II.

Increased visibility

A two-gun derivative featuring a single-bay wing design, a 110kW (150hp) Benz Bz.III engine and dispensing with the strut bracing for the tailplane, the D.IV, was rejected for service due to poor forward view. The Halberstadt D.V reverted to the basic design, single

---

Halberstadt D.II

Weight: (gross) 728kg (1606lb)

Dimensions: Length 7.3m (23ft 11in) Wingspan 8.8m (28ft 10in) Height 2.66m (8ft 9in)

Powerplant: One 90kW (120hp) Mercedes D.II 6-cylinder water-cooled inline piston engine

Speed: 150km/h (93mph)

Range: 250km (160 miles)

Ceiling: 4000m (13,000ft)

Crew: 1

Armament: One 7.92mm (0.312in) LMG 08/15 'Spandau' machine gun

---

Halberstadt D.III

Hans von Keudell was one of the founding members of Jasta 1, where he initially flew a Fokker D.I before being assigned this clear doped Halberstadt D.III bearing the marking 'K' for Keudell. After scoring a total of 12 victories, von Keudell was shot down and killed in an Albatros D.III by Stuart Pratt flying a Nieuport two-seater of 46 Squadron, Royal Flying Corps, in early 1917.

machine gun and Argus engine of the D.III and combined it with improved ailerons and a different cabane and upper-wing centre section design, resulting in better handling and improved pilot view. Around 90 examples of the D.V were constructed in two batches from late 1916 until June 1917, and gained a reputation as a delight to fly. However, by the time the final D.Vs rolled off the production line the basic design was no longer competitive against the latest Allied types and the D.V was replaced in the West during the summer of 1917, the last 31 examples produced being supplied to Ottoman

Halberstadt D.III

Weight: (gross) 728kg (1606lb)
Dimensions: Length 7.3m (23ft 11in) Wingspan 8.8m (28ft 10in) Height 2.66m (8ft 9in)
Powerplant: One 90kW (120hp) Argus As.II 6-cylinder water-cooled inline piston engine
Speed: 150km/h (93mph)
Range: 250km (160 miles)
Ceiling: 4000m (13,000ft)
Crew: 1
Armament: One 7.92mm (0.312in) LMG 08/15 'Spandau' machine gun

forces. Halberstadt D.Vs continued in frontline service in Palestine until the war's end, where, to deal with the high temperatures encountered in this

theatre, additional radiators were fitted to the fuselage sides resulting in a reduction in performance.

Halberstadt D.V

Weight: (gross) 728kg (1606lb)
Dimensions: Length: 7.3m (23ft 11in) Wingspan 8.7m (28ft 6in) Height 2.50m (8ft 2in)
Powerplant: One 90kW (120hp) Argus As.II 6-cylinder water-cooled inline piston engine or 90kW (120hp) Mercedes D.II 6-cylinder water-cooled inline piston engine
Speed: 160km/h (99mph)
Range: 200km (125 miles)
Ceiling: 4000m (13,000ft)
Crew: 1
Armament: One 7.92mm (0.312in) LMG 08/15 'Spandau' machine gun

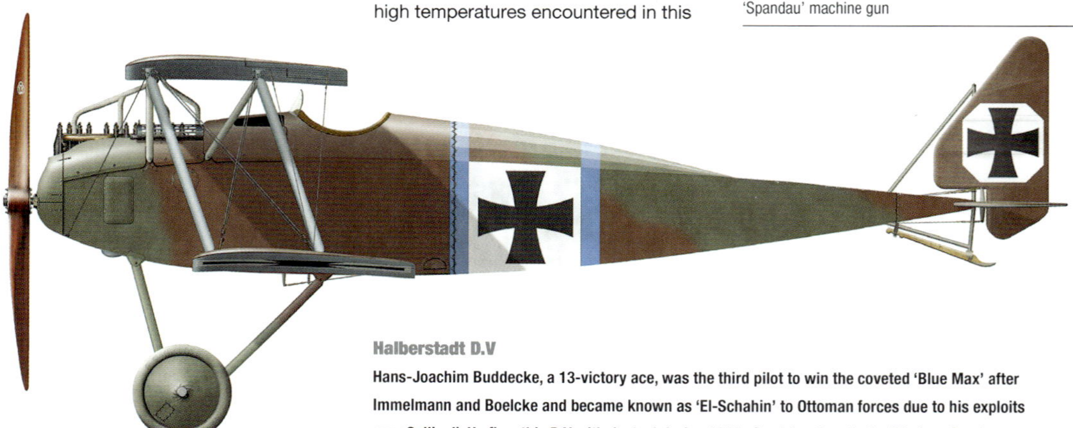

Halberstadt D.V

Hans-Joachim Buddecke, a 13-victory ace, was the third pilot to win the coveted 'Blue Max' after Immelmann and Boelcke and became known as 'El-Schahin' to Ottoman forces due to his exploits over Gallipoli. He flew this D.V with Jasta 4 during 1916 after his return to the Western Front.

Junkers D.I

Although its service life was virtually non-existent and fewer than 50 were ever built, the all-metal D.I was a revolutionary aircraft heralding a new vogue in aircraft design.

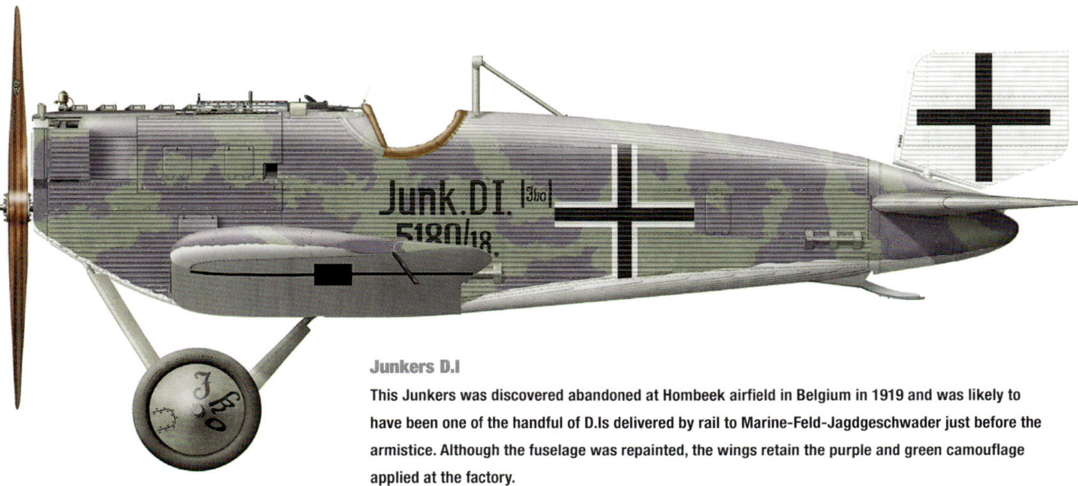

Junkers D.I

This Junkers was discovered abandoned at Hombeek airfield in Belgium in 1919 and was likely to have been one of the handful of D.Is delivered by rail to Marine-Feld-Jagdgeschwader just before the armistice. Although the fuselage was repainted, the wings retain the purple and green camouflage applied at the factory.

Junkers had produced the world's first all-metal aircraft, the J.1, and followed it up with an amazingly advanced machine, the J.7. The new design was a fighter, an all-metal cantilever monoplane, and it was years ahead of its time. To add to the radical features, ailerons were discarded in favour of all moving wingtips and the radiator was mounted on top of the engine, seemingly as an afterthought, totally obscuring the pilot's view directly ahead. The prototype was modified, incorporating the radiator into the nose and replacing the rotating wingtips with conventional ailerons, and in this form it was present at the First Fighter Competition at Adlershof in early 1918. Although not an official entry (because it was a monoplane), the J.7 demonstrated the highest speed of any aircraft in attendance, which attracted official attention. After formal testing IdFlieg enquired about the possibility of building a production run of J.7s, but by this time an improved version, the J.9, had been developed and this was an entrant at the Second Fighter Competition in May.

Clumsy handling

Judged unsuitable for aerial combat due to its relatively sluggish manoeuvrability, it was suggested that the J.9 would make an effective dedicated balloon destroyer as it was very fast and extremely robust, and the aircraft was ordered into production as the D.I. Only 41 were completed, the last being delivered in February 1919, and a mere handful reached the Front. It is unclear whether it saw any wartime action, although at least one Allied pilot reported encountering one in the closing days of the conflict. In early 1919 the Geschwader Sachsenberg, commanded by Naval ace Gotthard Sachsenberg, operated D.Is and two-seat CL.Is to support Freikorps units fighting Soviet forces on Germany's Baltic borders. During the fighting, the robust construction of the D.I rendered it largely impervious to gunfire and fully vindicated this remarkable design.

Junkers D.I

Weight: (gross) 834kg (1834lb)

Dimensions: Length 7.25m (23ft 9in) Wingspan 9m (29ft 6in) Height 2.60m (8ft 6in)

Powerplant: One 138kW (185hp) BMW IIIa water-cooled 6-cylinder inline piston engine

Speed: 170km/h (106mph)

Endurance: 1 hour 30 minutes

Ceiling: 5000m (16,400ft)

Crew: 1

Armament: One 7.92mm (0.312in) LMG 08/15 'Spandau' machine gun

Roland D.II & D.IIa

Nicknamed 'Haifisch' (shark) due to its streamlined shape, the Roland D.II was fast and strong but demonstrated underwhelming manoeuvrability and handling.

Roland D.II

The Roland D.II and D.IIa left the factory in a distinctive reddish-brown and green camouflage. Aircraft built by the LFG parent company are easy to distinguish by their 'fat' national markings, as applied to Fieseler's D.IIa pictured below. This Pfalz-built D.II has more conventionally proportioned markings and served with Jasta 32 on the Western Front.

With the success of its beautifully streamlined C.II two-seater, Luftfahrzeug Gesellschaft sought to apply the same formula, with its *wickelrumpf* fuselage construction to a single-seat fighter. The resulting aircraft was powered by a 119kW (160hp) Mercedes D.III engine and featured a distinctive streamlined pylon filling the gap between the fuselage and top wing, an aerodynamically efficient design solution but one that seriously impeded the pilot's view directly forward. Designated Roland D.I, approximately 20 aircraft had been delivered when a severe fire at the factory disrupted production. When it resumed, the aircraft featured a wing-mounted radiator in place of the fuselage 'ear' radiators, a tail skid fairing and an aerodynamically cleaned-up centre section and in this form it was designated D.II. Later examples fitted with the Argus As.III engine were dubbed D.IIa.

Limited appeal

Entering service during early 1917, the D.II was not met with universal enthusiasm, pilots criticising the heavy controls, the appalling forward view that made landing particularly difficult and the fact that manoeuvrability was worse than the Albatros D.III, an aircraft not itself noted for sparkling agility. Nevertheless, the aircraft was fast and sturdy, and although the Albatros remained the fighter of choice, several pilots scored victories with the Roland D.II and D.IIa before its withdrawal from combat units in December 1917. In total 300 examples of the D.II and D.IIa were built, most by Pfalz, who subsequently used the wickelrumpf construction technique

Roland D.IIa

Weight: (gross) 932kg (2055lb)

Dimensions: Length 6.8m (22ft 4in) Wingspan 8.9m (29ft 2in) Height 2.9m (9ft 6in)

Powerplant: One 119kW (160 hp) Mercedes D.III 6-cylinder water-cooled inline piston engine

Speed: 180km/h (112mph)

Endurance: 2 hours

Ceiling: 5000m (16,000ft)

Crew: 1

Armament: Two 7.92mm (0.312in) LMG 08/15 'Spandau' machine guns

Sleek but flawed, the Roland D.II nonetheless achieved a measure of success in less demanding combat arenas. This particular D.IIa was one of the substantial number of Roland fighters built by Pfalz Flugzeugwerke, who subsequently employed Roland's advanced construction techniques for their own, more potent fighter designs.

in all their own fighter designs. A final variant, the D.III, dispensed with the problematic pylon mounting of the top wing and replaced it with conventional cabane struts. Although view was much improved, the aircraft was inferior to contemporary Albatros and Pfalz designs and very few were built, nine being recorded at the Front in February 1918. A handful of D.IIs and D.IIIs were also supplied to the Bulgarian Army Aeroplane Section.

Roland D.IIa

Flying over the Macedonian Front, Gerhard Fieseler scored his first aerial victory in this Roland D.IIa on 20 August 1917. With 19 victories in total, Feiseler was the highest scoring ace on the Eastern Front to survive the war, but is better known today for the aircraft company that he founded to build his own designs, notably the Fieseler Fi 156 Storch.

Roland D.VIa & D.VIb

Although it had lost out to the all-conquering Fokker D.VII, the LFG Roland D.VI was ordered into production as an insurance against potential production problems with the Fokker aircraft.

Roland D.VIa

Otto Kissenberth was a remarkable ace: despite wearing glasses, he flew as a reconnaissance pilot from 1914 before transferring to single-seaters two years later. His personal edelweiss symbol graced the all-black fuselage of his Albatros D.V and this Roland D.VIa when his unit re-equipped with these aircraft in the spring of 1918. Unusually, Kissenberth scored the last of his 20 victories in a captured Sopwith Camel in May 1918.

Luft-Fahrzeug-Gesellschaft dropped their *wickelrumpf* system of fuselage construction in favour of *klinkerrumpf* (literally 'clinker built'), wherein the fuselage was formed by overlapping thin spruce strips over a light wooden framework much like a boat. As well as possessing great strength this method of construction allowed for a fine streamlined shape to be produced, although it was time-consuming to build and the D.VI was more expensive than the rival Fokker D.VII. The Roland first flew in late 1917 and the two prototypes were entered into the First Fighter Competition at Adlershof in January 1918. Although it failed to win, its performance was judged sufficiently good enough for production to be ordered by IdFlieg, with deliveries beginning in May.

Improved performance

Despite being judged inferior to the Fokker, the Mercedes D.III powered D.VI was still an excellent aircraft, offering roughly the same performance as the Albatros D.V but with superior manoeuvrability and devoid of the structural concerns of the Albatros fighter. The initial production version was the D.VIa that benefitted from the marginally more powerful Mercedes D.IIIa. In stark contrast to its earlier D.II, the view from the cockpit was very good and the Roland was further improved by the installation of a Benz Bz.IIIa engine of greater power – in this form it was designated D.VIb.

Late comer

The first D.VIs reached combat units during the late spring of 1918, with most being supplied to the Bavarian Jastas. Naval units tasked with protecting seaplane bases were also equipped with Roland D.VIs.

Never a common aircraft, 70 were recorded at the Front in August and production would continue until the armistice. After 1918, war-booty examples of the Roland went into service with the United States Army Air Corps and the D.VI formed the fighter equipment of the nascent Cezchoslovak Air force.

Roland D.VIb

Weight: (gross) 846kg (1865lb)

Dimensions: Length 6.32m (20ft 9in) Wingspan 9.42m (30ft 11in) Height 2.8m (9ft 2in)

Powerplant: One 149kW (200hp) Benz Bz.IIIa 6-cylinder water-cooled inline piston engine

Speed: 182km/h (114mph)

Endurance: 2 hours

Ceiling: 5800m (19,000ft)

Crew: 1

Armament: Two 7.92mm (0.312in) LMG 08/15 'Spandau' machine guns

Roland D.VIb

This attractively marked D.VIb is believed to have been operated with Jasta 51 and is one of the first D.VIbs to have been manufactured. It is also believed to have been one of the aircraft present at the Second Fighter Competition held at Adlershof in June and July of 1918.

Pfalz D.III & D.IIIa

A beautifully streamlined aircraft, the Pfalz D.III never quite lived up to its initial promise although it gave good if unspectacular service from the summer of 1917 to the end of hostilities.

Pfalz D.III

Resplendent in the blue and red of Jasta 5, this Pfalz D.III bears the personal markings of 44-victory ace Rudolf Berthold, one of the few significant pilots who expressed a positive opinion of the Pfalz. Berthold's right hand was paralyzed due to injury and the Pfalz's twist-grip throttle control on the joystick allowed him to overcome this impediment.

For the first two years of war the Pfalz company had largely concerned itself with the construction of other company's designs. In November 1916 however, the arrival of Austrian designer Rudolph Gehringer at Pfalz sparked development of the company's first original fighter design, the D.III. First flown in April 1917, the D.III was heavily influenced by the Roland D.II that Pfalz had been building immediately prior to the new fighter's debut. The new Pfalz design's performance was so obviously superior to that of the Roland that outstanding contracts held by Pfalz for the Roland D.II were amended in favour of the new aircraft and in June a further 300 were ordered.

Similar to the Albatros D.III and D.V, the Pfalz D.III was designed as a response to the Nieuport 17 that had proved superior to any fighter the Central Powers then possessed. It did not ape the Nieuport's sesquiplane layout to the same extent as the Albatros, and as a result never suffered from the structural concerns that bedevilled its more numerous contemporary. The D.III utilised Roland's patented *wickelrumpf* semi-monocoque fuselage construction that conferred great strength for light weight and allowed for the use of complex compound curves when compared to more conventionally constructed aircraft – although it was also time consuming and expensive to produce. The D.III was powered by a 120kW (160hp) Mercedes engine and armament consisted of two 7.92mm (0.312in) MG 08/15 machine guns, the standard for German fighters by 1917. These weapons were fitted within the fuselage so as not to interfere with the streamlined shape of the aircraft, an

Pfalz D.IIIa

Weight: (Maximum take-off) 933kg (2056lb)
Dimensions: Length 6.95m (22ft 10in) Wingspan 9.4m (30ft 10in) Height 2.67m (8ft 9in)
Powerplant: One 130kW (180hp) Mercedes D.IIIa 6-cylinder water-cooled inline piston engine
Speed: 165km/h (102mph)
Range: 400km
Ceiling: 6000m (19,685ft)
Crew: 1
Armament: Two 7.92mm (0.312in) LMG 08/15 'Spandau' machine guns

unusual feature that was carried over from Roland designs.

In-flight hazards

Performance on paper was broadly similar to the Albatros but the Pfalz was variously criticised for its comparatively poor manoeuvrability, slow rate of climb, heavy controls and vicious stall that led to a dangerous flat

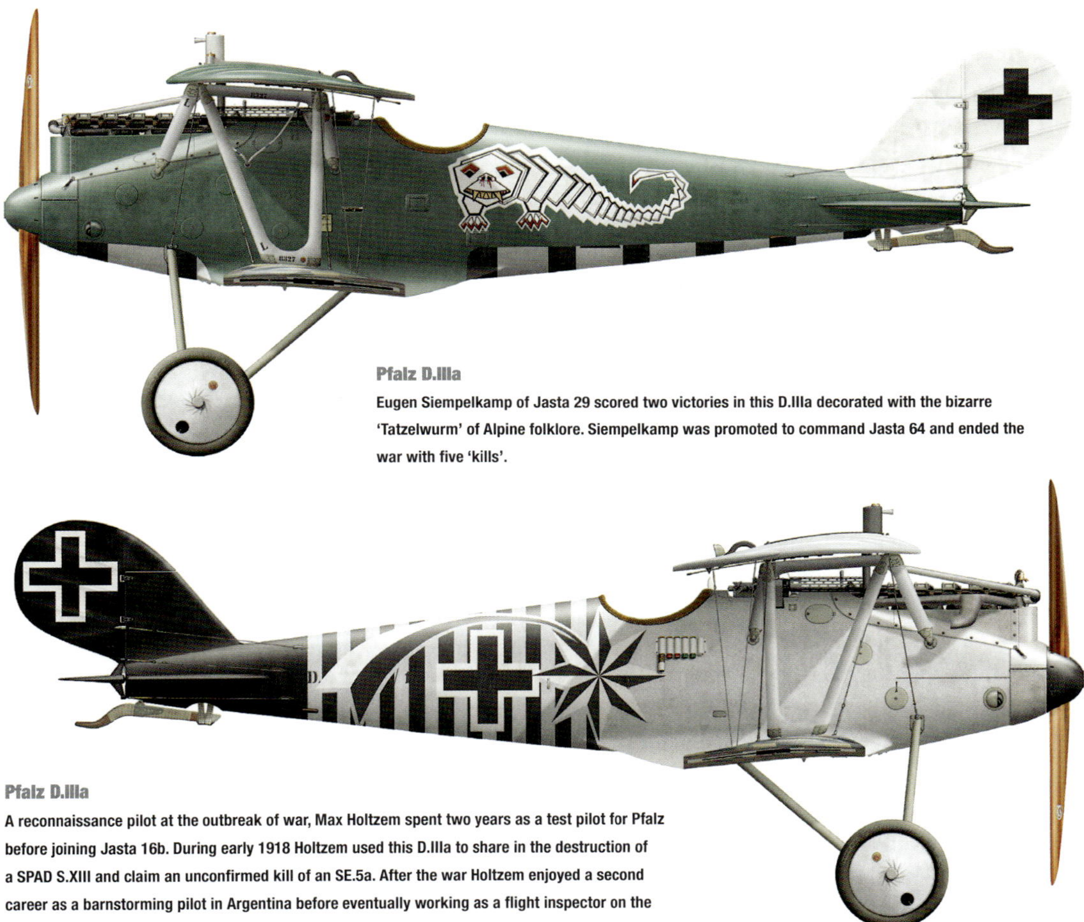

Pfalz D.IIIa

Eugen Siempelkamp of Jasta 29 scored two victories in this D.IIIa decorated with the bizarre 'Tatzelwurm' of Alpine folklore. Siempelkamp was promoted to command Jasta 64 and ended the war with five 'kills'.

Pfalz D.IIIa

A reconnaissance pilot at the outbreak of war, Max Holtzem spent two years as a test pilot for Pfalz before joining Jasta 16b. During early 1918 Holtzem used this D.IIIa to share in the destruction of a SPAD S.XIII and claim an unconfirmed kill of an SE.5a. After the war Holtzem enjoyed a second career as a barnstorming pilot in Argentina before eventually working as a flight inspector on the P-51 Mustang production line at North American.

spin. On the other hand, the Pfalz was very sturdily constructed, possessed excellent diving characteristics and could be flown to the limits of its performance with impunity. The most serious operational problem was the location of its machine guns in the fuselage; in the event of the guns jamming – a common occurrence at the time – the pilot was unable to gain access to the weapons to clear them. To answer this concern, in November 1917 Pfalz flew the improved D.IIIa in which the guns were relocated to an accessible position directly ahead of the pilot. The new version also featured rounded lower wing tips and enlarged horizontal tail surfaces to improve handling and about three-quarters of the approximately 1000 built were of the D.IIIa variant.

In use, the Pfalz D.III and D.IIIa gave good service although they were always overshadowed by contemporary Albatros fighters and somewhat outperformed by contemporary Allied fighters. Most pilots preferred the Albatros with its somewhat better handling and climb performance, and today the Pfalz D.III remains relatively obscure. Part of the reason for this was due to the fact that Pfalz was a Bavarian company and most (although not all) of its aircraft went to Bavarian units, which tended to be on quieter sections of the Front. After the introduction of the superlative Fokker D.VII, combat use of the D.III and D.IIIa declined – around 100 remained in front-line use at the armistice with several hundred more serving in advanced training units.

Pfalz D.VIII

An excellent aircraft blessed with a colossal rate of climb, production of the Pfalz D.VIII was restricted by the limited availability of the Siemens Halske Sh.III engine.

Pfalz D.VIII

178/18 was the personal machine of Ludwig Beckmann, commander of Jasta 56 who scored eight victories during the war and survived to command a Ju 52/3m transport unit during World War II. Jasta 56 generally sported yellow noses and tails but Beckmann, a Westphalian, adopted red on his own aircraft as red and white are the Westphalian national colours.

First flown in early 1918, the Pfalz D.VIII was a two-bay version of the slightly earlier single-bay D.VII design. Both were present at the First Fighter Competition at Adlershof in February 1918. Declared joint winner with the formidable Fokker D.VII, IdFlieg favoured the higher strength for slightly reduced performance offered by the D.VIII and a production order for 120 Pfalz D.VIIIs was placed. The relatively small size of the order reflected the relative scarcity of the Siemens-Halske engine, which was experiencing severe production problems.

Home-defence role

Examples of the D.VIII began to arrive at the Front in April 1918 but these were not released for service, probably due to ongoing engine issues, and it was June before any Pfalz D.VIIIs arrived at a combat unit. Subsequent service proved the Pfalz to be a formidable aircraft, the equal of the more widely-produced Siemens-Schuckert D.III and D.IV in climb performance but not as manoeuvrable. Therefore it was recommended that the D.VIII be employed as a home-defence interceptor and a few aircraft were indeed assigned to these units. Believed to have been operated by at least three Jastas over the Western Front, details of the D.VIII's service are scarce but it is known to have been used by 43-victory ace Paul Baumer and Harald Aufferth, who scored 30 kills. How many of those were achieved with the D.VIII remains unknown. Only around 40 D.VIIIs were produced and few reached combat units, though its performance was clearly excellent and it was lucky indeed for Allied aircrew that the aircraft was not available earlier in 1918.

Pfalz D.VIII

Weight: 738kg (1627lb)
Dimensions: Length 5.65m (18ft 6in) Wingspan 7.52m (24ft 8in) Height 2.75m (9ft)
Powerplant: One 119kW (160hp) Siemens-Halske Sh.III 11-cylinder air-cooled geared rotary engine
Speed: 190km/h (120mph)
Endurance: 1 hour 30 minutes
Ceiling: 6000m (19,685ft)
Crew: 1
Armament: Two 7.92mm (0.312in) LMG 08/15 'Spandau' machine guns

Pfalz D.XII

Produced to supplement the relatively scarce numbers of the outstanding Fokker D.VII, the Pfalz D.XII was an excellent fighter that never escaped the shadow of its more famous contemporary.

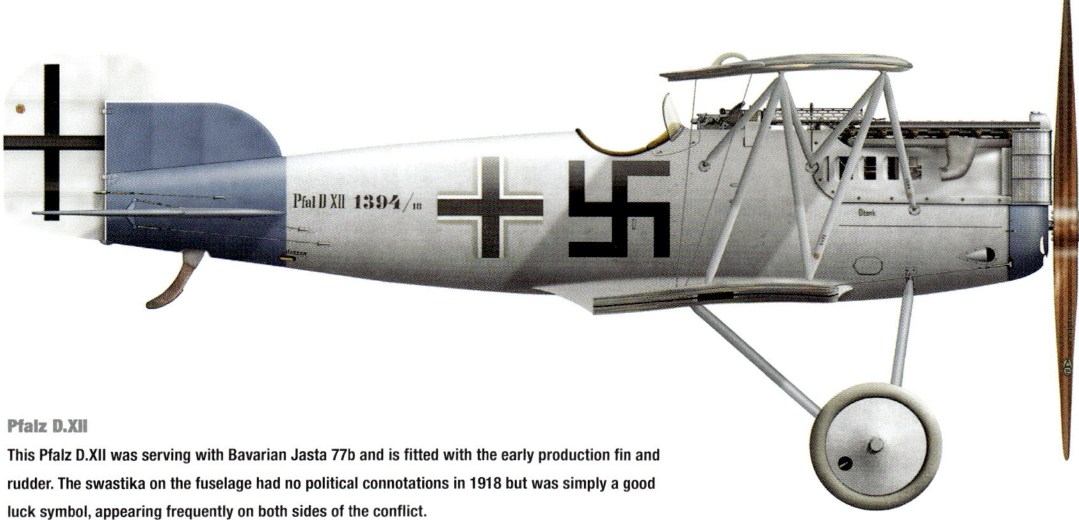

Pfalz D.XII

This Pfalz D.XII was serving with Bavarian Jasta 77b and is fitted with the early production fin and rudder. The swastika on the fuselage had no political connotations in 1918 but was simply a good luck symbol, appearing frequently on both sides of the conflict.

All of the Pfalz fighter designs to see production were destined to play second fiddle to a more popular rival. The Pfalz Eindeckers were considered inferior to the Fokkers, the Pfalz D.III was regarded as an understudy to the Albatros D.III and the D.XII would never gain the accolades lavished on the Fokker D.VII. In the latter case this does seem to have been rather unfair as the D.XII was a highly capable aircraft.

Designer Rudolf Goeringer sought to maintain the excellent strength characteristics of Pfalz's D.IIIa while improving overall performance. To this end Goeringer reduced the overall dimensions of the fuselage, which once again utilised the wickelrumpf construction system to impart great strength and a smooth streamlined shape, and combined it with a new set of wings. In a marked contrast to the Fokker D.VII's thick cantilever units, the Pfalz mounted a two-bay set of very thin, heavily braced wings known to the Germans as 'SPAD-type' due to their resemblance to those of the very successful SPAD S.VII. This arrangement proved to be light but did contribute to drag. Flown for the first time in April 1918, various engines were tested but the Mercedes D.IIIa was selected, all production of the superior BMW being allocated to the Fokker D.VII programme.

'Clumsy cart-horse'

Entered into the Second Fighter Competition at Adlershof in June 1918, the Pfalz D.XII put up a good showing but most pilots considered

Pfalz D.XII

Weight: (gross) 897kg (1978lb)

Dimensions: Length 6.3m (20ft 10in) Wingspan 9m (29ft 6in) Height 2.7m (8ft 10in)

Powerplant: One 130kW (180hp) Mercedes D.IIIa 6-cylinder liquid-cooled inline piston engine

Speed: 170km/h (110mph)

Endurance: 2 hours 30 minutes

Ceiling: 5600m (18,500ft)

Crew: 1

Armament: Two 7.92mm (0.312in) LMG 08/15 'Spandau' machine guns

it inferior to the D.VII, though leading aces Ernst Udet and Hans Weiss rated it highly (leading to subsequent unproved rumours that both were bribed by the Pfalz company). Performance, on paper at least, was roughly equivalent to the Fokker, the D.XII possessing a higher diving speed despite its less powerful engine, but in reality it was heavier on the controls and lacked the docile handling characteristics of the Fokker. Rudolf Stark, commander of Jasta 35b, would later write, "The Pfalz was a clumsy cart-horse that went heavy in the reins and obeyed nothing but the most brutal force."

Production

Partly as a result of the influential Udet's good opinion but also as a consequence of the Kingdom of Bavaria's separate procurement policy (Pfalz being a Bavarian company), the D.XII entered large-scale production with the first examples reaching frontline units in July 1918, the majority unsurprisingly being supplied to Bavarian Jastas. Exact production figures are unknown, but orders were placed for at least 750 aircraft, with around 550 delivered by the armistice.

Unpopular model

The Pfalz D.XII was not particularly popular with service pilots despite offering a useful increase in performance over the Albatros D.V and with none of the structural concerns of the earlier aircraft, or the suspicious build quality bedevilling Fokker products. Unfortunately it was regarded as deficient in manoeuvrability, particularly in roll rate, and like most contemporary aircraft it had an abrupt stall and fell easily into a spin. Furthermore, its complicated rigging took a long time to set up correctly and the aircraft was unpopular with ground crews. Nonetheless, against the latest Allied aircraft the D.XII proved effective and after adjusting to its handling qualities, many pilots grew to appreciate its qualities – some even liked it. Ultimately, the Pfalz D.XII's greatest problem wasn't any inherent flaw in the aircraft itself, but simply that it wasn't a Fokker D.VII.

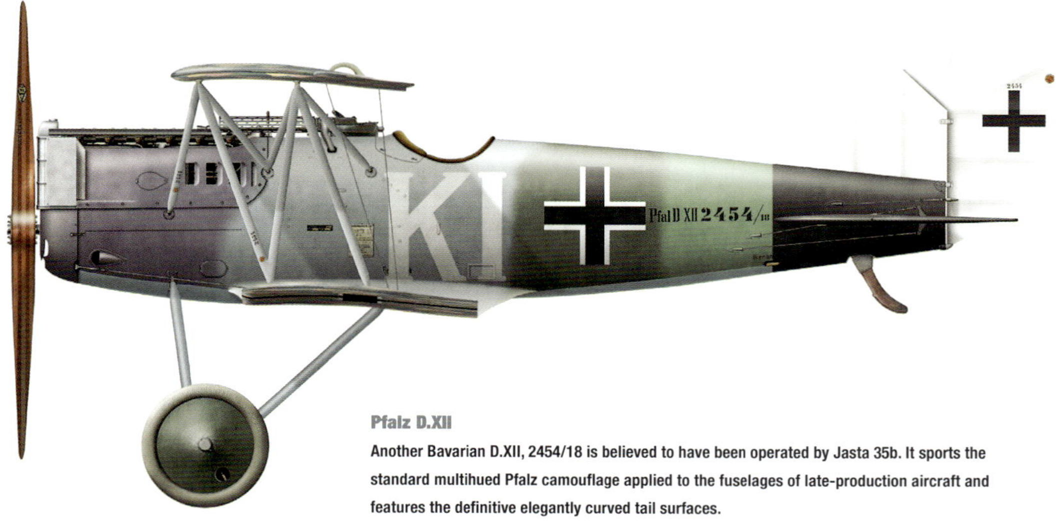

Pfalz D.XII

Another Bavarian D.XII, 2454/18 is believed to have been operated by Jasta 35b. It sports the standard multihued Pfalz camouflage applied to the fuselages of late-production aircraft and features the definitive elegantly curved tail surfaces.

Siemens-Schuckert D.I

Almost unique in being a reverse-engineered example of an enemy aircraft, the Siemens-Schuckert Werke (SSW) D.I was unable to match the spectacular success of its progenitor.

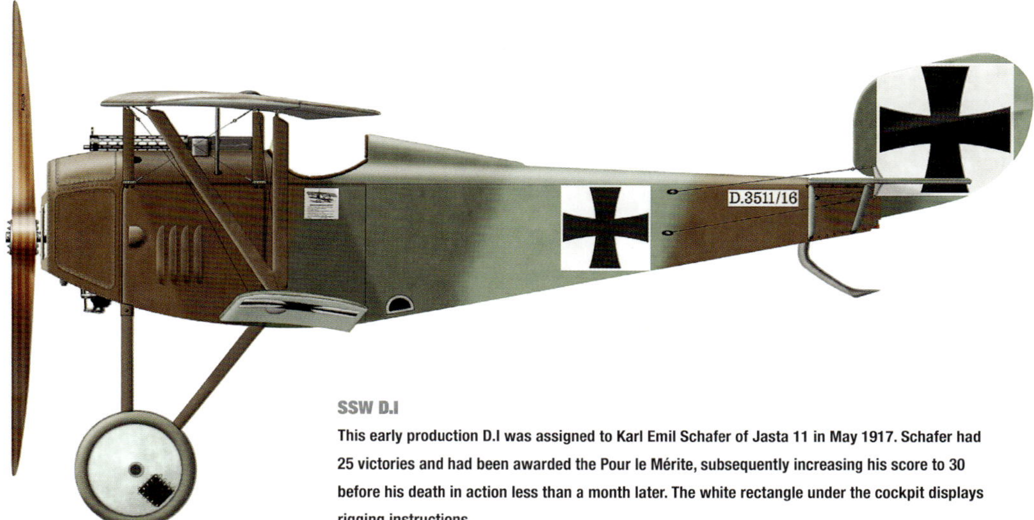

SSW D.I

This early production D.I was assigned to Karl Emil Schafer of Jasta 11 in May 1917. Schafer had 25 victories and had been awarded the Pour le Mérite, subsequently increasing his score to 30 before his death in action less than a month later. The white rectangle under the cockpit displays rigging instructions.

It is difficult to overestimate the effect the diminutive Nieuport 17 had on German fighter development. During the summer of 1916 the Nieuport demonstrated marked superiority over the Fokker and Pfalz monoplanes and captured examples of the French fighter were supplied to several manufacturers. Its design greatly influenced a crop of German fighters but IdFlieg also contracted with the Siemens-Schuckert and Euler companies to produce a reverse-engineered copy of the Nieuport, reasoning that this would significantly reduce development time and allow a capable fighter to arrive at the Front promptly, pending the arrival of more potent domestic designs.

Euler produced a handful of Nieuport clones that were almost exclusively used for training due to the inadequate power of their engines, but the Siemens-Schuckert D.I was rather more successful, orders for some 350 being placed by March 1917. Although almost a direct copy of the Nieuport, the D.I did differ in some respects, most significantly in its 82kW (110hp) Siemens-Halske Sh.I engine. The German engine rotated more slowly than the Le Rhône 9Ja of the original, requiring a propeller of greater diameter to be fitted, and the D.I sported a taller undercarriage to allow clearance for the larger airscrew. The gap between the upper wing and fuselage was also reduced and most were fitted with a prominent propeller spinner.

Although a delight to fly, the D.I's performance was inferior to the original Nieuport 17. Slow engine deliveries severely delayed production, by which time sufficient Albatros D.I and D.IIs with their superior performance and

Siemens-Schuckert D.I

Weight: (gross) 675kg (1488lb)
Dimensions: Length 6m (19ft 8in) Wingspan 7.5m (24ft 7in) Height 2.6m (8ft 6in)
Powerplant: One 82kW (110hp) Siemens-Halske Sh.I 9-cylinder air-cooled rotary piston engine
Speed: 155km/h (96mph)
Endurance: 2 hours 20 minutes
Ceiling: 5300m (17,400ft)
Crew: 1
Armament: One 7.92mm (0.312in) LMG 08/15 'Spandau' machine gun

armament had reached the Front. Production of the Siemens-Schuckert D.I was terminated at the 95th example in July 1917 although a further 55 partially completed examples were supplied for training aviation mechanics. Its modest performance resulted in a brief career in the West, and most operational D.Is were delivered piecemeal to various Jastas on the Eastern Front.

Siemens-Schuckert D.III

Despite its rotund appearance, the Siemens-Schuckert D.III was one of the most potent fighters of the entire conflict. Possessed of an outstanding rate of climb, the D.III was likely the finest interceptor to see widespread service before the armistice.

In anticipation of a new 11-cylinder rotary engine of unprecedented power from the affiliated Siemens-Halske manufacturer, Siemens-Schuckert constructed three prototype aircraft of the same basic design but differing in detail. Designated the D.II, D.IIa and D.IIb, the first two airframes were completed in January 1917. Unfortunately the complicated Siemens-Halske Sh.III was severely delayed and nearly six months was to pass until the first flight-ready engine became available for testing and the aircraft made its first flight on 7 June 1917. Outright speed of the prototypes was not outstanding but climb performance was remarkable. On 5 August company test pilot and 12-victory ace Hans Muller took the D.IIb to an altitude of 7000m (22,965ft) in 35.5 minutes, which was an unofficial record, receiving a bonus of 1500 Marks in the process. Three further development aircraft were completed in October and a pre-production run of 20 D.IIIs was placed in December.

Landing problems

In January the First Fighter Competition was held at Adlershof and Siemens-Schuckert entered three prototypes and one pre-production D.III. Despite Muller demonstrating the outstanding climb performance of the aircraft, service pilots found the controls too heavy, manoeuvrability lacking, and considered the D.III too difficult to land. Five-kill ace Hans von der Osten compared the D.III to the Fokker D.VII prototype as "An elephant to a mosquito." Ultimately, three of the aircraft crashed in landing

Siemens-Schuckert D.III

Weight: (gross) 725kg (1598lb)
Dimensions: Length 5.7m (18ft 8in) Wingspan 8.43m (27ft 8in) Height 2.8m (9ft 2in)
Powerplant: One 120kW (160hp) Siemens-Halske Sh.III 11-cylinder air-cooled geared rotary piston engine
Speed: 177km/h (110mph)
Range: 360km (220 miles)
Ceiling: 8000m (26,000ft)
Crew: 1
Armament: Two 7.92mm (0.312in) LMG 08/15 'Spandau' machine guns

SSW D.III

Joachim von Ziegesar flew this early D.III with Jasta 15. Note the fully cowled engine, differently shaped rudder and lack of cooling vents in the larger spinner when compared with the later example of Ernst Udet pictured opposite. Ziegesar scored three confirmed victories, survived the war, and during the 1930s wrote a thrilling account of flying the Siemens-Schuckert in action.

SSW D.III

Ernst Udet, Germany's most successful surviving ace, was photographed in this aircraft when in command of Jasta 4 at Metz in September 1918, although it is unknown if he ever flew it in combat. Like all Udet's aircraft of 1917 and 1918, it carries the pet name 'Lo' of his fiancee Elanore Zink.

accidents during the competition, although all the pilots escaped relatively unharmed. Siemens undertook alterations during the course of the competition, greatly improving handling, and adopted a four-blade propeller allowing for a shorter undercarriage, significantly lessening landing problems. The D.III had already demonstrated excellent altitude performance but its speed below 2000m (6560ft) was also found to be slightly better than the Fokker D.VII and although the latter aircraft emerged as the overall winner, IdFlieg ordered 80 D.IIIs in March 1918 as a result of the competition results.

In total, 41 D.IIIs were delivered to combat units by May 1918, most going to Jasta 2 under Rudolf Berthold, where they proved immediately effective, Berthold noting the D.III's 'brilliant' rate of climb. However, all was not well with the Sh.III engine and catastrophic failures were occurring after only seven to 10 hours' running. The problem was the lack of castor oil or adequate alternative lubricant, the synthetic replacement Voltol being found to lack sufficient quality, a problem exacerbated by a consignment of the lubricant of the wrong viscosity being delivered in error for use in the D.III. Operations were paused during May while the D.IIIs were returned to the factory for engine replacement, IdFlieg taking the opportunity to demand a new rudder, larger cockpit, different propeller, aileron and elevator improvements, and a cutaway cowling to improve cooling. By July the revised Sh.IIIa engine had successfully completed its 40-hour reliability test and the aircraft was returned to service.

Front-line instruction

Although very manoeuvrable, it was noted that the D.III required a higher degree of piloting skill than the docile Fokker D.VII. As a result two Siemens-Schuckert test pilots, Hans Muller and Bruno Rodschinka, toured front-line units to instruct service pilots and perform flying demonstrations, Muller being present on 21 August when Naval ace Theo Osterkamp intercepted a DH.4 at 6000m (19,685ft). The combined effects of these visits and initial combat victories succeeded in instilling confidence in the type, such that a pilot of Kest 8 wrote, "the Siemens is much superior to Allied aircraft" on 2 October. The D.III served with great effect until the end of hostilities.

Siemens-Schuckert D.IV

Taking the already impressive performance of the D.III to its wartime zenith, the Siemens-Schuckert D.IV was one of very few aircraft that could genuinely rival the Fokker D.VII as the best German fighter of the war.

It came as a surprise to both Siemens-Schuckert and the officials of IdFlieg that service pilots present at the First Fighter Competition unanimously regarded maximum speed as the greatest single attribute of any single-seat fighter. As a result, Siemens-Schuckert reworked its D.III with wings of shorter chord and lesser area, sacrificing rate of climb for greater speed. In this form the D.IV, as it had been designated, was entered into the Second Fighter Competition of May and June 1918. In the opinion of the front-line pilots present, the D.IV was the finest of the Sh.IIIa-powered fighters, more manoeuvrable than the second placed Pfalz D.VIII. As a result it was decided that the Pfalz would be recommended for home-defence duties and the Siemens-Schuckert

would be assigned to the Front. IdFlieg placed an order for 60 D.IVs on 26 July, with follow-on orders for a further 250 placed by the armistice, although only 123 are known to have been completed during the war, around half of which made it to combat units.

Treaty of Versailles

In front-line units the D.IV proved popular and effective. Even Jagdgeschwader Richthofen, who had rejected the D.III, after a demonstration by Hans Muller on 5 October requested, "twelve good D. IV fighters and a further twelve as soon as the first are shipped". Performance was fantastic, climb rate was still excellent, and above 4000m (13,120ft) it was both faster and more manoeuvrable than the vaunted Fokker D.VII. Production even continued

after the armistice, during which time Siemens-Schuckert developed the sesquiplane D.V and the extremely promising D.VI parasol monoplane before the Treaty of Versailles halted all German military aircraft development and production in June 1919.

Siemens-Schuckert D.IV

Weight: (gross) 735kg (1620lb)
Dimensions: Length 5.7m (18ft 8in) Wingspan 8.35m (27ft 5in) Height 2.72m (8ft 11in)
Powerplant: One 120kW (160hp) Siemens-Halske Sh.IIIa 11-cylinder air-cooled geared rotary piston engine
Speed: 190km/h (120mph)
Range: 380km (240 miles)
Ceiling: 8000m (26,000ft)
Crew: 1
Armament: Two 7.92mm (0.312in) LMG 08/15 'Spandau' machine guns

---

SSW D.IV

Although precious few D.IVs reached the Front, those that did performed brilliantly. One of the pilots lucky enough to fly it was Hermann Becker, commander of Jasta 12. He scored his 23rd and last victory in this machine when he downed a SPAD XIII on the afternoon of 3 November 1918.

Fokker Dr.I

Inextricably connected with Manfred von Richthofen, the Fokker triplane remains the most famous aircraft of the Great War. In reality the Dr.I was at best a qualified success: structurally suspect and demanding to fly, but possessed of incredible agility.

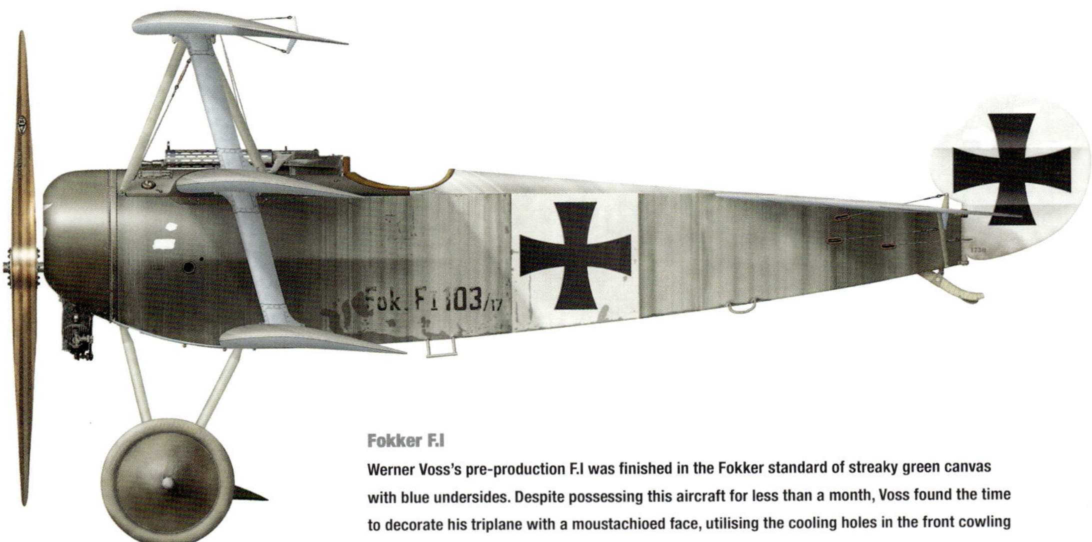

Fokker F.I

Werner Voss's pre-production F.I was finished in the Fokker standard of streaky green canvas with blue undersides. Despite possessing this aircraft for less than a month, Voss found the time to decorate his triplane with a moustachioed face, utilising the cooling holes in the front cowling for eyes.

Throughout World War I German authorities showed a much greater willingness to encourage the incorporation of enemy fighter developments into their own designs than their French or British counterparts. The Nieuport 17's sesquiplane layout influenced a whole generation of German fighters, and slightly later the spectacular success of the Sopwith Triplane inspired a veritable frenzy of triplane designs to appear from several German manufacturers, but only the Fokker Dr.I would be produced in any numbers. Fokker followed the triplane layout from the Sopwith but in all other regards the aircraft was an entirely original design, smaller in overall dimensions and featuring three cantilever wings,

this being obvious on the first prototype, the Fokker V.4, which featured no interplane struts. First flown in the early summer of 1917, the prototype was reworked into the V.5, of which three were built. Designated F.I by IdFlieg, two of these pre-production prototypes saw operational service. One was flown by Richthofen to score his 60th victory but was later destroyed in combat with Sopwith Camels, costing the life of 33-victory ace Kurt Wolff, and the other was flown by the mercurial Werner Voss.

High praise

Voss was at this time second only to Richthofen in overall victories and he was extremely enthusiastic about the F.I, recommending its general adoption after test-flying one of the brand-new

Fokker Dr.I

Weight: (gross) 586kg (1291lb)

Dimensions: Length 5.77m (18ft 11in) Wingspan 7.19m (23ft 7in) Height 2.95m (9ft 8in)

Powerplant: One 82kW (110hp) Oberursel Ur.II 9-cylinder air-cooled rotary piston engine

Speed: 180km/h (110mph)

Range: 300km (190 miles)

Ceiling: 6100m (20,000ft)

Crew: 1

Armament: Two 7.92mm (0.312in) LMG 08/15 'Spandau' machine guns

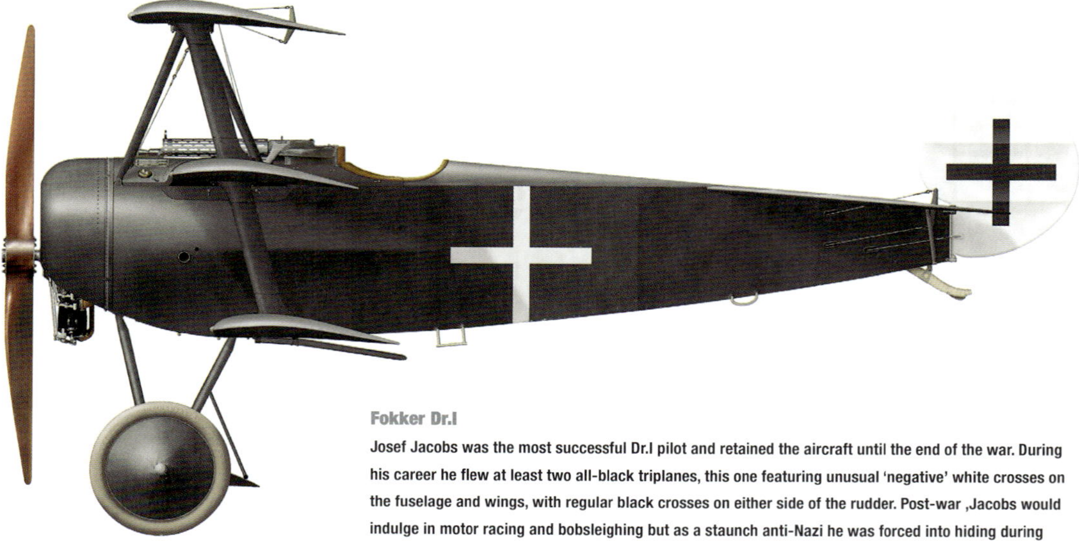

Fokker Dr.I

Josef Jacobs was the most successful Dr.I pilot and retained the aircraft until the end of the war. During his career he flew at least two all-black triplanes, this one featuring unusual 'negative' white crosses on the fuselage and wings, with regular black crosses on either side of the rudder. Post-war ,Jacobs would indulge in motor racing and bobsleighing but as a staunch anti-Nazi he was forced into hiding during World War II.

triplanes at Fokker's Schwerin factory in July 1917. In August this F.I was assigned to Voss as his personal aircraft and he would score his final 11 kills in it before losing his life in a legendary dogfight when he single-handedly fought six SE.5as, all of which were flown by aces of the elite 56 squadron RFC, managing to hit each of his attackers with gunfire until, after 10 minutes of combat, his engine failed for unknown reasons and he was shot down. Noting the enthusiasm of these outstanding pilots for the new aircraft, the F.I went into production as the Fokker Dr.I (the 'Dr' standing for Dreidecker: 'triplane'). Differing from the F.I only in possessing horizontal tail surfaces of a slightly different shape (the leading edge of the tailplane was straight rather than slightly curved as on the F.I), and featuring skids under the wingtips to prevent damage in the event of a ground loop, IdFlieg ordered 100 Dr.Is

in September and a further 200 in November.

Initial impressions of the Dr.I were good: the aircraft was outstandingly manoeuvrable, especially compared to the Albatros and Pfalz fighters then in use, and possessed an excellent rate of climb. Manfred von Richthofen noted that It could "climb like a monkey and manoeuvre like a devil".

Impressive handling

Handling the aircraft was not considered particularly difficult but the instability of the Dr.I meant that it was impossible to fly 'hands-off' for any length of time and the aircraft effectively had to be constantly controlled by the pilot. Skilled airmen found this invigorating, Franz Hemer – who would later fly the demonstrably superior Fokker D.VII – wrote: "The triplane was my favourite fighting machine because it had such wonderful flying qualities. I could let myself stunt –

looping and rolling – and could avoid an enemy by diving with perfect safety." Roll rate was not particularly good due to the relatively ineffective ailerons fitted to the top wing only, but the aircraft possessed light and powerful rudder and elevators and could be yawed to a remarkable degree and remain controllable.

This quality was used to great effect by Voss in his final flight and was remarked upon by SE.5a pilot Geoffrey Bowman: "To my amazement he kicked on full rudder, without bank, pulled his nose up slightly, gave me a burst while he was skidding sideways and then kicked on opposite rudder before the results of this amazing stunt appeared to have any effect on the controllability of his machine." Like its great adversary the Sopwith Camel, the torque of the rotary engine meant that turns to the right in the Dr.I were quicker than those to the left.

Grounded

The Dr.I seemed to vindicate its designers and was generally regarded as the best fighter available to the Jastas in the autumn of 1917. Unfortunately all was not well with the diminutive triplane. Heinrich Gontermann, a 39-victory ace, was killed when his Dr.I broke up in mid-air as he performed aerobatics over his airfield. Two days later Gunther Pastor of Richthofen's Jasta 11 suffered the same fate, this time in level flight. IdFlieg grounded the triplane (only 17 were in service at the time) and halted production pending investigation. Tests revealed that the wings of the triplanes were poorly constructed and insufficient waterproofing was causing the structure to fail. Anthony Fokker was ordered to replace the wings on all existing Dr.Is at his own expense and build future wings to a higher standard. The triplane was returned to

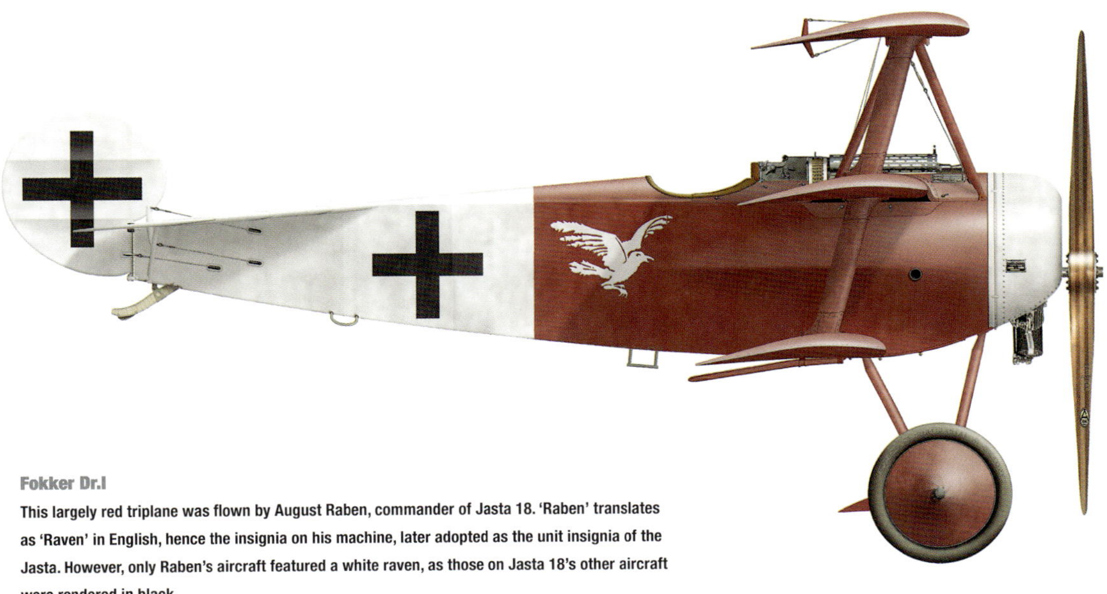

Fokker Dr.I

This largely red triplane was flown by August Raben, commander of Jasta 18. 'Raben' translates as 'Raven' in English, hence the insignia on his machine, later adopted as the unit insignia of the Jasta. However, only Raben's aircraft featured a white raven, as those on Jasta 18's other aircraft were rendered in black.

Fokker Dr.I

One of a number of Dr.Is flown by the 'Red Baron', it was serial number 425/17 in which von Richthofen met his death. It is uncertain whether he fell victim to a Sopwith Camel flown by Arthur Brown or by ground fire from Australian troops.

Tail

The 'comma' rudder without a separate fin was a classic trademark of Anthony Fokker's early fighter designs. A tailskid was fitted immediately below the rudder.

Lifting surface

The impressive agility of the Fokker Dr.I was further enhanced by an additional lifting surface (aerofoil) that enclosed the axle of the fixed main undercarriage.

Fokker Dr.I

Weight: (gross) 586kg (1291lb)

Dimensions: Length 5.77m (18ft 11in) Wingspan 7.19m (23ft 7in) Height 2.95m (9ft 8in)

Powerplant: One 82kW (110hp) Oberursel Ur.II 9-cylinder air-cooled rotary piston engine

Speed: 180km/h (110mph)

Range: 300km (190 miles)

Ceiling: 6100m (20,000ft)

Crew: 1

Armament: Two 7.92mm (0.312in) LMG 08/15 'Spandau' machine guns

Gun armament

The twin LMG 08/15 'Spandau' guns were arranged side-by-side in the upper part of the forward fuselage. Air-cooled and belt-fed, the weapons were each provided with 500 rounds of ammunition, housed behind the fuel tank.

Configuration

In its three-wing layout, the Dr.I was one of a number of contemporary fighting scouts to adopt the proven configuration of the British Sopwith Triplane.

operations by the end of November and production resumed in late December. Despite the improvement in quality control, failures of the upper wings persisted. Hans Joachim Wolff and Lothar von Richthofen (Manfred's younger brother), both of Jasta 11, managed to survive upper-wing failures in February and March of 1918 respectively and it was this unfortunate tendency that primarily contributed to the termination of Dr.I production in May 1918 after the construction of a mere 320 examples.

'Wing and a prayer'

Post-war testing using equipment unavailable at the time revealed that the upper wing was generating up to two-and-a-half times more lift than the other two, and was literally trying to tear itself off the aircraft at all times. Gradually withdrawn from front-line Jastas, the Dr.Is were reassigned to training and home-defence units. In the training role the Dr.I was often fitted with the lower powered 75kW (100hp) Goebel Goe.II

in place of the Oberursel. A few of the more influential pilots retained a Dr.I for their personal use and thus a few persisted in service at the Front until the armistice – the last known example being that of Josef Jacobs, the most successful Dr.I exponent with 24 of his approximately 48 victories achieved with the triplane, scoring the last on 28 October 1918.

It is curious that an aircraft that was produced in such limited numbers and saw widespread service for a relatively short period of time should be so famous today. The reason, of course, is its association with the Red Baron. Even during the short period he flew the triplane, Manfred von Richthofen and his Dr.I had been subject to assiduous propaganda that ascribed a certain mystique to the aircraft.

Richthofen achieved most of his victories in Albatros aircraft but at the height of his fame he flew the triplane almost exclusively, shooting down 19 of his 80 total victories with it until his death in his red-painted Dr.I on

21 April 1918. One of the triplanes Richthofen flew actually survived the war, to be put on display at the Zeughaus Museum in Berlin, but it was destroyed by British bombing in 1943, an unfortunate end for this undeniably charismatic machine.

THE RED BARON
Born into an aristocratic family in Silesia in 1892, Manfred Freiherr von Richthofen was the leading air ace of World War I, with 80 aerial victories. Commencing flying training in May 1915, he began his career as an observer on the Eastern Front, before training as a fighter pilot. Transferring to the Western Front, and Jasta 2, he scored his first kill near Cambrai on 17 September 1916, his victim an F.E.2b. After 16 victories in Albatros D.IIs, von Richthofen was given command of Jasta 11, where he scored 40 kills in a period of just six months. He was appointed commander of the first Jagdgeschwader in June 1917, and he was leading JG 1 when killed in action on 21 April 1918.

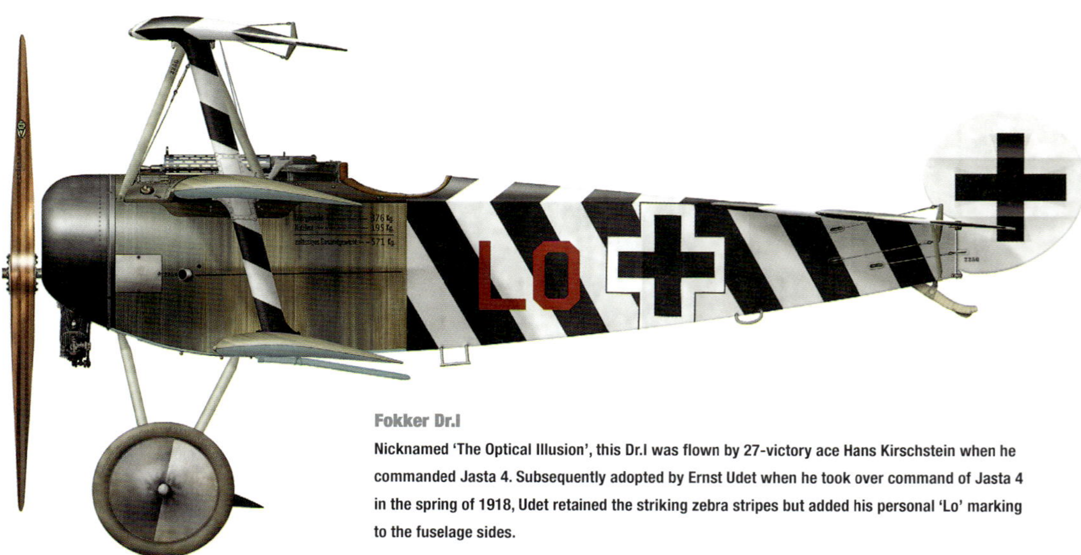

Fokker Dr.I

Nicknamed 'The Optical Illusion', this Dr.I was flown by 27-victory ace Hans Kirschstein when he commanded Jasta 4. Subsequently adopted by Ernst Udet when he took over command of Jasta 4 in the spring of 1918, Udet retained the striking zebra stripes but added his personal 'Lo' marking to the fuselage sides.

Werner Voss was a great friend of Manfred von Richthofen, an exceptional pilot, and an enthusiastic exponent of the Fokker triplane. This F.I pre-production aircraft (illustrated on page 47) was assigned to Voss for his personal use. The face derives from designs on Japanese fighting kites and Voss had the aircraft fitted with a French Le Rhône engine in place of the standard Oberursel.

Fokker Eindecker

The first aircraft to enter service equipped with an interrupter gear allowing the machine gun to fire through the propeller, Fokker's E series of monoplanes was the world's first effective fighter and one of the most significant combat aircraft in history.

The Germans lagged behind the Allies in the adoption of machine gun armament to their aircraft. By early 1915, both French and British 'pusher' two-seater aircraft were regularly armed with light machine guns but German forces possessed no such weapon. The appearance therefore of the armed Eindecker during the summer of 1915 was not only a profound shock to the Allies but also represented an amazing reversal in German technical capability. This was only possible due to the development of the Parabellum MG 14 and Bergmann MG, which slowly became available for aviation units in early 1915.

Synchronizing fire

The other major technological development introduced by the Eindecker was a reliable means to fire the machine gun so that the bullets passed between the propeller blades. French aviator Roland Garros had fitted triangular metal plates to the blades of his Morane-Saulnier Type L monoplane to deflect the occasional bullet that would otherwise have severed the propeller.

Unfortunately for Garros, after achieving the first three air-to-air victories by firing a machine gun through the propeller disc, he was brought down in German territory by ground fire. Despite setting fire to his damaged aircraft, the gun and propeller remained intact and the German military demanded a similar solution. Brilliant Dutch designer Anthony Fokker would later claim that he developed the synchronization gear

Fokker E.I

Weight: (gross) 550kg (1212 lb)
Dimensions: Length 6.95m (22ft 8in) Wingspan 8.95m (29ft 4in) Height 2.9m (9ft 6in)
Powerplant: One 60kW (80hp) Oberursel U.0 9-cylinder air-cooled rotary piston engine
Speed: 130km/h (81mph)
Endurance: 1 hour 30 minutes
Ceiling: 3000m (9800ft)
Crew: 1
Armament: One 7.92mm (0.312in) Parabellum MG 14 or one 7.92mm IMG 08 'Spandau' machine gun

Fokker E.I

A Fokker E.I 5/15 was used by Kurt Wintgens to score the first Eindecker victory, thus beginning the 'Fokker Scourge' and ushering in a new era in air combat. The aircraft is depicted with the black rudder of FFA 48, with whom Wintgens was serving when he achieved his first confirmed kill in July 1915. The object protruding from the cockpit is a headrest, initially deemed necessary for the pilot to use the gunsight effectively.

Fokker E.II

Baron Kurt von Crailsheim transferred to aviation after being wounded in the infantry in August of 1914. Crailsheim flew this E.II with FAA 53 during the autumn of 1915, replacing an earlier Eindecker that he had written off in a crash landing following engine failure – a common occurrence at this early stage of military aviation development.

used on the Eindecker within 48 hours of being made aware of it by German high command, but as with many of Fokker's more grandiose claims, this was an outright lie as Fokker had been working on just such a system for at least six months.

Indeed, a patent for a synchronizing system was issued to Franz Schneider as early as July 1913, full details of the system being published in the German magazine Flugsport in September 1914 but the Prussian War Ministry had shown no interest in the idea. The pressures of war had served to change that position.

Fokker fitted a synchronized Parabellum MG 14 to an example of his M.5K monoplane, essentially an improved copy of a Morane-Saulnier type G, and demonstrated it to German officials in mid-May 1915. Impressed with the results, five Fokker A.IIIs were fitted with guns and redesignated the Fokker E, the 'E' standing for 'Eindecker mit MG': 'monoplane with machine gun'. Two pilots, Otto Parschau and Kurt Wintgens, were particularly closely involved with evaluating the armed

M.5Ks on operations, and on 1 July 1915 Wintgens shot down a Morane-Saulnier Type L, the first victory by a German fighter aircraft and the first anywhere for an aircraft using a synchronized machine gun. This and a second Morane downed three days later had to remain unconfirmed as the aircraft both fell on the French side of the lines, but on 15 July Wintgens scored a confirmed victory over a third Morane, this becoming the first confirmed victory by a Fokker Eindecker. Two further pilots who would soon gain considerable fame – Oswald Boelcke and Max Immelmann – were issued with two of the first five M.5K Eindeckers during the same month and both would gain victories during August.

The immediate success of the Fokker led to a production order and ultimately 54 examples of the E.I would be built. Production E.Is featured the wing lowered slightly from the shoulder of the fuselage to its midpoint to improve pilot view, and standardised on the Maschinengewehr IMG 08 machine gun after the synchronised Parabellum proved seriously unreliable.

Fokker E.II

Weight: (gross) 604kg (1330lb)

Dimensions: Length 7.22m (23ft 8in) Wingspan 8.85m (29ft) Height 2.8m (9ft 2in)

Powerplant: One 75kW (100hp) Oberursel U.I 9-cylinder air-cooled rotary piston engine

Speed: 140km/h (87mph)

Endurance: 1 hour 30 minutes

Ceiling: 3000m (9800ft)

Crew: 1

Armament: One 7.92mm (0.312in) IMG 08 'Spandau' machine gun

Engine availability

Before the E.I had even seen service Fokker had designed and built the E.II. Fitted with the 75kW (100hp) Oberursel U.I, a copy of the French Gnome Monosoupape, in place of the E.I's 60kW (80hp) Oberursel U.0, itself a copy of the Gnome Lambda; both variants were produced concurrently. The decision whether a particular airframe was finished as an E.I or E.II depended solely on engine availability. The E.II also featured wings of reduced span in an attempt to increase speed, although handling and climb performance suffered, and a slightly lengthened fuselage to balance the increased weight of the more powerful engine.

'Fokker Fodder'

In total, 49 E.IIs had been built by the time the factory switched to the main production variant, the E.III, in December 1915. By this time the Eindecker pilots of the Fleigertruppe were scoring freely and it is difficult to overstate the effect this was having on both sides. To the British, the phrase the 'Fokker Scourge' soon caught the popular imagination after being coined by the press in mid-1916, and the phrase 'Fokker Fodder' was used to describe British pilots – questions even being asked in Parliament, so serious was the situation perceived to be. In hindsight, this hysteria seems somewhat bizarre.

Actual losses to the Fokkers was light by later standards, with a total of 28 victories being claimed by Eindecker pilots for the final six months of 1915, but the perception of loss of air superiority to a supposedly invincible German aircraft had a catastrophic effect on morale: British pilot and author Cecil Lewis recounted: "Hearsay and a few lucky encounters had made the machine respected, not to say dreaded by the slow, unwieldy machines then used by us." Meanwhile, in Germany Boelcke and Immelmann were lionised in the Press, much to the shy Boelcke's embarrassment, and as early as the spring of 1916 Immelmann's E.I had been selected for preservation in the Saxon Army Museum, demonstrating the reverence with which the Eindecker was regarded.

Fokker E.III

Weight: (gross) 610 kg (1345 lb)
Dimensions: Length 7.25m (23ft 9in) Wingspan 10.05m (32ft 11in) Height 2.4m (7ft 10in)
Powerplant: One 75kW (100hp) Oberursel U.I 9-cylinder air-cooled rotary piston engine
Speed: 140km/h (87mph)
Endurance: 2 hours 30 minutes
Ceiling: 3600m (11,810ft)
Crew: 1
Armament: One 7.92mm (0.312in) LMG 08/15 'Spandau' machine gun

Fokker E.III

Ernst Udet became Germany's most successful fighter pilot to survive the war, with 62 confirmed victories. The first of these he achieved in this Fokker E.III serial number 105/15 on 18 March 1916 when at 5.10 PM he shot down a French Farman over Muhlhausen.

Fokker E.IV

Max Immelmann was the most famous of the Eindecker pilots, scoring all 15 of his victories with Fokker monoplanes. As well as giving his name to an aerial manoeuvre known as the 'Immelmann Turn,' Prussia's highest military honour, the *Pour le Mérite*, was allegedly nicknamed the 'Blue Max' in his honour – Immelmann being the first recipient of this medal during the Great War. Immelmann used this E.IV 127/16 to score three confirmed and one unconfirmed victory in March 1916.

E.III

By this time the E.III was serving at the front in numbers. Fitted with the same engine as the E.II, the E.III differed only in that it was fitted with new wings of a slightly narrower chord and possessed a larger fuel tank allowing for a two-and-a-half hour flight endurance. In total 249 would be built and a number of the E.IIs were converted to E.III standard when they were returned to Fokker's Schwerin factory for repair. By now, however, newer Allied aircraft – namely the D.H.2 and Nieuport sesquiplanes – possessed performance considerably in excess of the Eindecker, and the capture of an intact E.III in April 1916 revealed to surprised Allied authorities just how modest the Fokker monoplane's performance actually was. Meanwhile, the Nieuport in particular encouraged a reaction amongst German aircrew similar to that engendered by their own Eindeckers mere months earlier.

E.IV

Many pilots took to flying captured Nieuports in combat and IdFlieg took the drastic step of ordering direct Nieuport copies from Siemens-Schuckert and Euler. Seeking to wrest back air superiority, Fokker developed the E.IV, which featured the unprecedented armament of three LMG 08 'Spandau' machine guns all firing through the propeller, although synchronisation failures led to two such guns being a more common fitment.

Powered by the 120kW (160hp) Oberursel U.III two-row rotary, the E.IV was easily distinguishable by the 'turtledeck' fairing around the cockpit. The powerful U.III engine bestowed the E.IV with a comparatively impressive turn of speed, but its weight and the large gyroscopic forces associated with such a large spinning mass combined with the archaic wing-warping lateral control meant the E.IV was significantly less nimble than its predecessors. Furthermore, the U.III was extremely unreliable, tending to lose power after only a few hours of operation.

Only 49 examples of the E.IV were delivered, the last in December 1916, by which time the Eindecker was thoroughly obsolete.

Fokker E.IV

Weight: (gross) 724kg (1596lb)

Dimensions: Length 7.5m (24ft 7in) Wingspan 10m (32ft 10in), Height: 2.7m (8ft 10in)

Powerplant: One 120kW (160hp) Oberursel U.III 14-cylinder air-cooled rotary piston engine

Speed: 170km/h (110mph)

Endurance: 1 hour 30 minutes

Ceiling: 3960m (12,990ft)

Crew: 1

Armament: Two or three 7.92mm (0.312in) LMG 08/15 'Spandau' machine guns

---

Pfalz E-types

Contemporaries of the successful Fokker Eindeckers, the Pfalz monoplanes were regarded as inferior by their crews but offered similar performance and contributed to the initial success of the German fighters.

At the start of the war the Bavarian Pfalz company was licence building French Morane-Saulnier aircraft and many unarmed examples of Pfalz-built Morane monoplanes were used by German forces in the first two years of war. Following the demonstration of Anthony Fokker's gun synchronisation equipment, it was decided to try fitting the armament to the Pfalz monoplane and this was deemed suitably successful for production to commence during 1915 as the Pfalz E.I, of which around 45 were constructed.

This was followed by 130 examples of the E.II, which mounted a 75kW (100hp) Oberursel U.I in place of the 60kW (80hp) U.O of the E.I, but was otherwise identical. The E.III was an armed version of the Morane 'Parasol' but only four reached the Front.

A switch to the considerably more powerful, although unreliable, 120kW (160hp) 14-cylinder two-row Oberursel U.III rotary resulted in the Pfalz E.IV, but the improvement in performance was not as great as anticipated and production halted after 46 were built. The E.V saw a switch to the six-cylinder inline Mercedes D.I but by this time the Pfalz airframe was obsolete and only 20 were built.

Fokker scourge

The Pfalz was never received with the same enthusiasm as its Fokker contemporaries, and although it saw fairly widespread service its performance and manoeuvrability never quite matched its rival. Better built than the Fokker, its wooden structure was basically unchanged from the pre-war Morane design and was not sufficiently strong enough to deal with the stresses of increasingly high-G aerial combat.

With the appearance of superior biplane fighters, the operational life of the Pfalz monoplane in the West was short, but it saw considerably longer service over the Eastern Front.

Ironically, due to its near identical appearance to the Fokker monoplane, the Pfalz E-types contributed significantly to the notion of the 'Fokker Scourge' and the near mythic status of Fokker's Eindecker.

Pfalz E.I

Weight: (maximum take-off) 535kg (1177lb)

Dimensions: Length 6.3m (20ft 8in) Wingspan 9.26m (30ft 5in) Height 2.55m (8ft 4in)

Powerplant: One 60kW (80hp) Oberursel U.O 9-cylinder air-cooled rotary piston engine

Speed: 145km/h (91mph)

Endurance: 2 hours

Ceiling: 3000m (9800ft)

Crew: 1

Armament: One 7.92mm (0.312in) LMG 08/15 'Spandau' machine gun

---

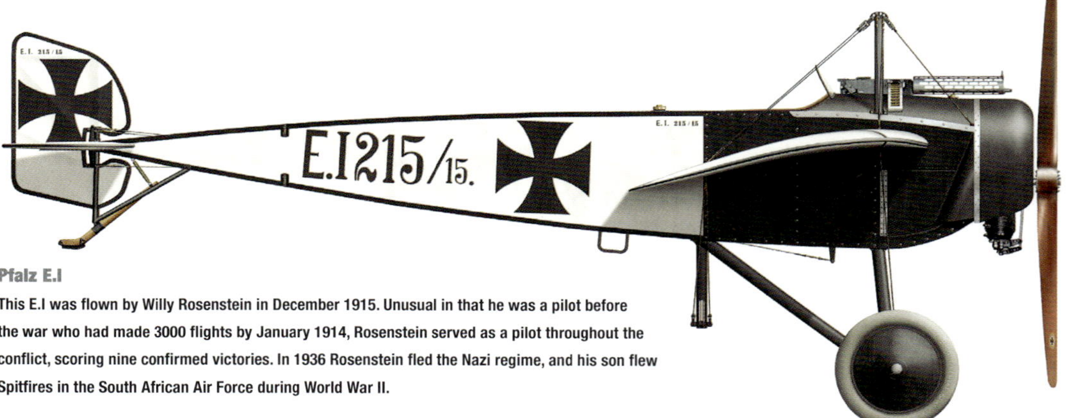

Pfalz E.I

This E.I was flown by Willy Rosenstein in December 1915. Unusual in that he was a pilot before the war who had made 3000 flights by January 1914, Rosenstein served as a pilot throughout the conflict, scoring nine confirmed victories. In 1936 Rosenstein fled the Nazi regime, and his son flew Spitfires in the South African Air Force during World War II.

Judging by the raised tail and the fact that the aircraft is rigidly supported under its struts rather than on its wheels, this immaculate Pfalz E.I was probably having its machine gun bore-sighted, a process regularly undertaken to maintain the accuracy of the sights. The Pfalz was never as popular as the Fokker but achieved good results in combat.

Pfalz E.II

This E.II can be distinguished from the E.I by its slightly longer chord cowling. The Pfalz E-types' black-edged fuselage led to their morbid nickname of 'flying death notice' as contemporary newspapers printed death notices with a similar border. Kurt Jentsch, who flew this E.II, remarked that the, "Pfalz monoplane was about as well suited for air combat as a cow was for playing the lute."

Pfalz E.II

Weight: (maximum take-off) 620kg (1364lb)
Dimensions: Length 6.45m (21ft 2in) Wingspan 10.2m (33ft 6in) Height 2.55m (8ft 4in)
Powerplant: One 75kW (100hp) Oberursel U.I 9-cylinder air-cooled rotary piston engine
Speed: 150km/h (94mph)
Endurance: 2 hours
Ceiling: 3600m (11,810ft)
Crew: 1
Armament: One 7.92mm (0.312in) LMG 08/15 'Spandau' machine gun

TWO-SEATER RECONNAISSANCE & GENERAL PURPOSE

Reconnaissance was by far the most important task of the German air services during the war and aircraft were initially perceived exclusively as observation tools for ground forces. As the war progressed so did the effectiveness of aerial reconnaissance. Aircraft were routinely equipped with radio to communicate aiming instructions to artillery, and long-range photoreconnaissance machines ranged far behind Allied lines.

This chapter includes the following aircraft:

- Albatros B-types
- Aviatik B-types
- Fokker A-types
- Taube
- AEG C-types
- Albatros CI. to C.III
- Albatros later C-types
- Aviatik C-types
- DFW C-types
- Halberstadt C.V
- LFG Roland C.II & C.IV
- LVG C.II
- LVG C.V & C.VI
- Rumpler C.I
- Rumpler C.III & C.IV

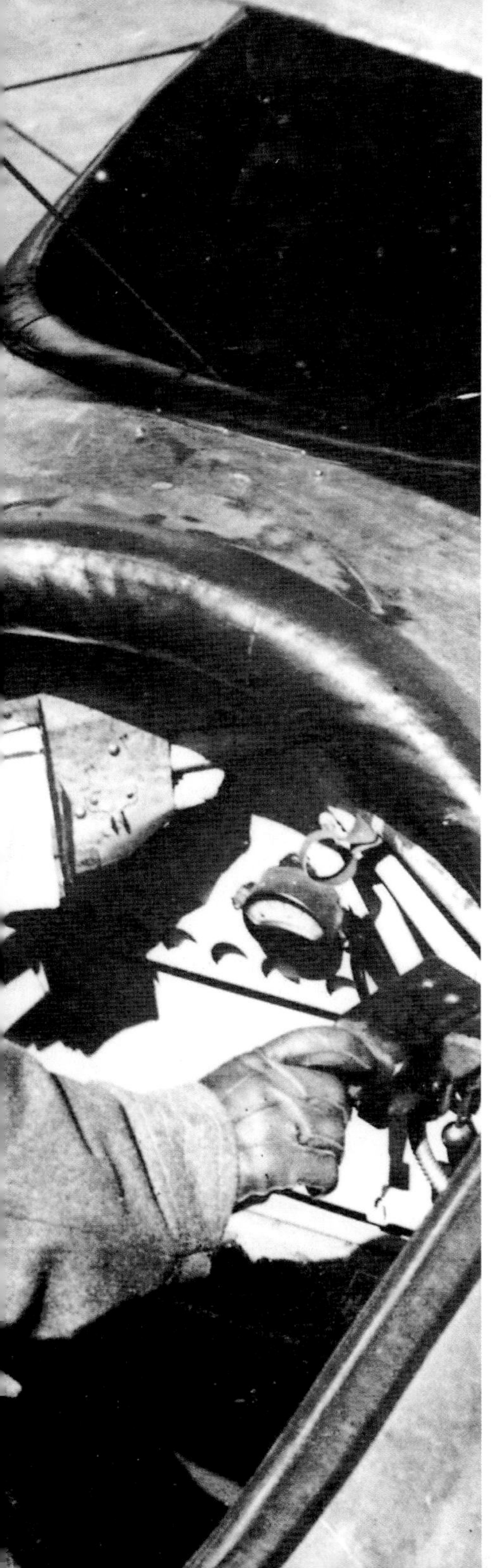

Warmly wrapped up in an army greatcoat, M1913 flying helmet and non-standard-issue woollen scarf, the observer of an AEG C.IV operates the Morse key of his aircraft's wireless. His 7.92mm Parabellum MG14 machine gun is mounted on a rail that runs around the sides and rear of the cockpit, allowing him to slide it into position when required.

Albatros B-types

The Albatros B-types were very successful in the early war period and the B.II subsequently enjoyed an astonishingly long career as a trainer.

Albatros developed two conventional two-seat biplanes concurrently during 1913, the DD (*Doppeldecker*) and the slightly longer DDK with a three-bay wing design. The latter aircraft entered service before the war as the B.I in the reconnaissance role during the war's opening weeks and enjoying a longer career as a training machine. Only a small number were constructed before production concentrated on the DD, designated B.II by IdFlieg, which featured a two-bay wing design of reduced span. Shortly before the outbreak of war a B.II set an altitude record of 4500m (14,764ft). Remarkably, the B.II would be manufactured by seven different companies in large numbers, although exact figures are unknown. Production continued for the duration of the war and beyond, the only major change being the substitution of all steel tube airframe components for wood due to shortages.

Training role

Initially employed as a reconnaissance and general purpose machine, the B.II achieved a certain notoriety by becoming the German Army's first aircraft to attack Britain. On 16 April 1915, the observer of a B.II from Feldflieger Abteilung 41 dropped 10 bombs, by hand, in the area around Sittingbourne in Kent, causing virtually no damage. From around mid-1915, the B.IIs were gradually replaced by armed C-type aircraft and the unarmed and slow B-types were relegated to training, a role in which the B.II excelled. As well as service with the Luftskreitkräfte and Austria-Hungary, the aircraft was operated by eight other nations including, surprisingly, the UK. The sole British example was purchased for evaluation in early 1914 and served until early 1918 with the RNAS. In Sweden the B.II was licence-built by several companies and the type was finally withdrawn in 1935.

Albatros B.II

Weight (gross) 1071kg (2361lb)

Dimensions Length 7.63m (25ft 0.38in) Wingspan 12.8m (42ft) Height 3.15m (10ft 4in)

Powerplant One 74.5kW (100hp) Mercedes D.I, or 89.5kW (120hp) Mercedes D.II, or 74.5kW (100hp) Argus As.1, or 89.5kW (120hp) Argus As.II, or 82kW (110hp) Benz Bz.II 6-cylinder water-cooled inline piston engine

Speed 120km/h (75mph)

Endurance: 4 hours

Ceiling: 3000m (9840ft)

Crew: 2

Armament: None

Albatros B.II

This early production B.II had a replacement rudder fitted replacing the original smaller unit and obliterating half the tailplane eisernkreuz in the process. As built, it would have mounted a radiator above the engine but has been retrofitted with fuselage mounted Hazet radiators. The engine in this example was the Mercedes D.II.

Aviatik B-types

The Aviatik B-types derived from a pre-war design and provided excellent service until the vulnerability of operating an unarmed aircraft over the Front became untenable.

Aviatik B.II

The Aviatik was a stalwart of early-war German aviation and was reliable and safe enough to provide a training platform for many years after its front-line career had ended. This B.II was serving with the Beobachterschule (Observers School) at Colonia-Butzweilerhof in 1916.

Automobil und Aviatik AG was established in 1909 and, after building a selection of French designs under licence, produced its first indigenous aircraft, the P.13, in 1912. Designed by Robert Wild and first flown in either April or May, the P.13 was produced in several variants, featuring three-and-a-half or four-bay wing designs. A developed version with detail improvements to the fuselage and a shorter two-and-a-half bay wing went into production in 1913 as the P.14, with a further improved version, the P.15, featuring a tail fin and two or three bay wings also appearing in 1913.

The Luftstreitkräfte ordered 101 P.13s and P.14s in 1913, an enormous order by contemporary standards. Designated simply Aviatik B, these aircraft were followed into service by the P.15, which was designated B.I and the P.15b, a lightly modified version with Mercedes D.II engine that became the B.II in September 1915. Further large orders followed and the aircraft gave yeoman service in the reconnaissance role, in the process earning the dubious honour of becoming the first aircraft to be shot down in aerial combat when an Aviatik B was downed on 5 October 1914 by a French Voisin III.

Foreign production

Unusually, the Aviatik B also saw large wartime production and service use on the Allied side. SAML of Italy had obtained a production licence and manufactured 410 to the original design as the S.I before switching to the machine gun equiped S.2, of which several hundred saw service with Italian units from 1917 onwards. Some examples of the SAML aircraft eventually made their way to Paraguay, where at least one was still serving as a trainer during the Chaco war of 1932–1935. This was a remarkable longevity of service for an aircraft whose basic design dated from 1912.

Aviatik B.I (P.15)

Weight: (gross) 1090kg (2403lb)

Dimensions: Length 7.63m (25ft) Wingspan 14m (45ft 11in) Height 3.05m (10ft)

Powerplant: One 74.5kW (100hp) Mercedes D.I 6-cylinder water-cooled inline piston engine

Speed: 100km/h (62mph)

Endurance: 4 hours

Ceiling: 5000m (16,404ft)

Crew: 2

Armament: None

Fokker A-types

Fokker produced many unarmed A-type monoplanes for observation and reconnaissance in the first two years of the war, one of which was developed into the armed M.5K 'Eindecker.'

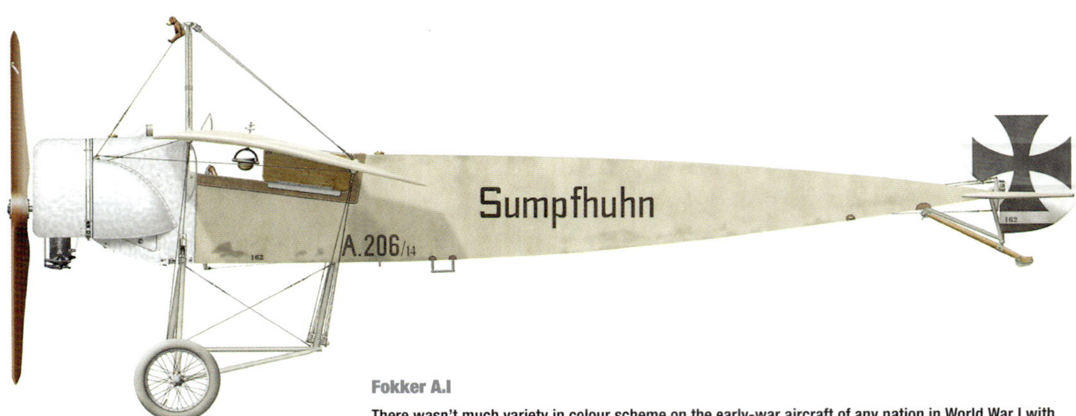

Fokker A.I

There wasn't much variety in colour scheme on the early-war aircraft of any nation in World War I with clear doped linen and bare metal being the norm. Considerations of camouflage or flamboyant personal decoration came later; however, the first inkling of personalization can be seen on this A.I, which has been named 'Sumpfhuhn', meaning 'Moorhen' in English.

Although best remembered today for giving rise to the spectacularly successful Eindecker fighter, Fokker's earlier monoplanes were important early war aircraft in their own right. Before the outbreak of war, Fokker had been engaged in producing the M.5, an unofficial copy of the Morane-Saulnier H with numerous improvements, not least a steel tube fuselage imparting greater strength than the original.

First into military service in numbers was the two-seat M.8, which was known as the A.I in Fliegertruppe service. A somewhat unusual design, the A.I was almost a parasol monoplane in its prototype form, with the leading edge of the wing suspended above the fuselage on struts but the rear spar attached to the upper longerons. Production aircraft featured a built-up forward fuselage decking to meet the front spar of the wing, resulting in a rather compromised view forwards although the fuselage sides were cut away either on side of the cockpit to provide downward visibility. Around 63 were built, many under subcontract by Halberstadt.

Fokker also produced a single-seat reconnaissance aircraft, the M.5, first flown in the spring of 1914, which appeared in two distinct versions, the M.5K (Kurz Spannweite – short wingspan), which became the Fokker A.III in military service, and the M.5L (Lange Spannweite – long wingspan), which became the A.II. Even before the war, several M.5s had been acquired privately by Prussian Army officers and two had been supplied to the Austro–Hungarian Army. The M.5 in both forms entered production in 1914, and the

Fokker A.I

Weight: (gross) unknown
Dimensions: Length 7.54m (24ft 9in) Wingspan 12.12m (39ft 9in) Height 2.75m (9ft)
Powerplant: One 60kW (80hp) Oberusel U.0 9-cylinder air-cooled rotary piston engine
Speed: 135km/h (84mph)
Endurance: 4.5 hours
Ceiling: 3000m (9800ft)
Crew: 2
Armament: None

aircraft proved popular with its pilots, possessing a sprightly performance for the time, although developments would very soon outstrip it.

In early 1915 five M.5Ks were ordered with a single Parabellum MG14 synchronised to fire through the propeller disc, these serving as the pre-production airframes for the E.I 'Eindecker' fighter.

Taube

Germany's most iconic aircraft of the pre-war period, the Taube ('Dove') had already made history as the first aircraft to drop a bomb in combat. Its operational career during World War I was brief but eventful.

Designed in 1909 by Austro-Hungarian Igo Etrich, the Taube's distinctive wing planform was based on the gliding Zanonia seed, which possesses inherent stability in both pitch and roll, a quality considered highly desirable for aircraft at the time. The Taube flew for the first time in 1910 with Etrich at the controls, although he passed on testing duties to pilot Karl Illner after he nearly broke his back when the Taube crashed on an early test flight. Despite this mishap, the aircraft was generally successful and Etrich tailored the designs to meet the specific criteria of the military that included such 'useful' attributes as the ability to land on a freshly ploughed field. Edmund Rumpler obtained a five-year licence to manufacture Taubes in July 1910 and five Rumpler Taubes were ordered by the Prussian Army after demonstrations during October, the

first of many to enter military service. Unfortunately for Etrich, the German patent office invalidated the Taube patent in September 1911 and the design became public. As a result no fewer than 14 manufacturers would ultimately build them, and differences abounded across the aircraft produced by these firms. The total number of Taubes built is not known with any accuracy today, but is believed to be around 500.

The basic Taube structure was unusual even by contemporary standards. Despite being a heavily rigged monoplane the Taube also possessed a truss structure consisting of a heavily cable braced steel tube underneath the wing to impart strength. Roll control was by wing warping, specifically the rearward projecting wingtips and there was no elevator as such, the rear of the

Jeannin Stahltaube

Weight: (gross) 1035kg (2020lb)
Dimensions: Length 9.69m (31ft 10in) Wingspan 13.87m (45ft 6in) Height 2.97m (9ft 9in)
Powerplant: One 89kW (120hp) Argus As.II 6-cylinder water-cooled inline piston engine
Speed: 100km/h (62mph)
Endurance: 4 hours
Ceiling: 2000m (6600ft)
Crew: 2
Armament: None

Jeannin Taube A.271

The largest producer of Taubes was Rumpler but Jeannin was responsible for one of the next most commonplace Taube clones. A.271 was at Adlershof for type testing by IdFlieg in early 1915 but appears to have been retained for communications general usage, as it was photographed there painted in these markings, which were introduced in October 1916.

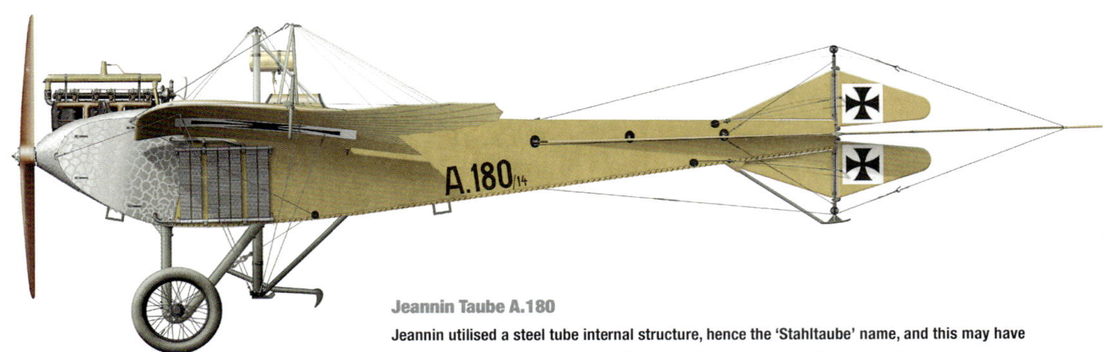

Jeannin Taube A.180

Jeannin utilised a steel tube internal structure, hence the 'Stahltaube' name, and this may have contributed to the longevity of A.180, the world's sole surviving Taube. Selected for museum preservation during World War I, it narrowly avoided destruction by Allied bombing in World War II and can be seen today on display at the Deutsches Technikmuseum, Berlin.

tailplane could be warped up or down to provide vertical control. The rudder, which was split with equal parts above and below the horizontal tail surfaces, was the only conventionally hinged control surface. Taubes were fitted with a variety of engines, mostly inline units, initially the four-cylinder Mercedes E4F, but Taubes from some manufacturers also utilised rotary engines. Initially the Taube was highly praised for its relative ease of control and reliability. In 1911 Helmut Hirth won the Kathreiner Prize for a flight from Munich to Berlin, some 540km (355 miles), which he achieved in around six hours (excluding fuel stops) – which was noteworthy at the time for being faster than the train. During December the same year Gino Linnekogel and Suvelick Johannisthal set an endurance record for a two-seater of four hours and 35 minutes.

Combat role

By this time the Taube had been used in combat, becoming the first aircraft to attack an enemy force when an Italian example piloted by Giulio Gavotti dropped four grenades on Ottoman forces during the Battle of Ain Zara, and Taubes were subsequently used to drop bombs during the First Balkan War of 1912–1913. In August 1914, all German civilian Taubes were commandeered by the military and many were employed during the opening months of the war.

Famously a single example became the first aircraft to drop a bomb on Paris on 30 August 1914 accompanied by a note stating: "The German army is at the gates of Paris. There is nothing for you to do but surrender," and in October a Taube bombed Japanese ships blockading the Chinese port of Tsingtao (although causing no damage).

Despite these aggressive missions, the vast majority of operations were for reconnaissance and observation – for example, Taubes were responsible for spotting advancing Russian troops at the Battle of Tannenberg, directly leading to the decisive German victory.

Unfortunately for the Taube's operational career, its great stability, low power and lack of structural strength rendered it extremely vulnerable and the aircraft was withdrawn from the Front after around six months. Its career as a reliable and docile training aircraft extended into 1916 and many of the most famous German pilots made their first flights in a Taube.

AEG C-types

Electrical giant AEG built a selection of mediocre two-seaters before developing the successful C.IV of 1916, which was built in large numbers and served until the end of the war.

AEG had produced an unarmed B-type general reconnaissance aircraft for the German air arm but with a maximum speed of a mere 110km/h (68mph), its performance was so pedestrian, even for 1914, that the few built were soon discarded. AEG used the aircraft as a basis to develop an armed C-type machine with a more powerful 112kW (150hp) Benz BZ.III in place of the Mercedes D.II of the B.II, the performance was sufficiently improved for serial production to go ahead, maximum speed being raised to 130km/h (81mph). Only a few C.Is were built before AEG switched in October 1915 to the smaller C.II featuring a slightly improved performance and the ability to carry a small bombload. The C.III featured a fuselage that filled the gap between the two wings and placed the gunner directly above the top wing, giving him an unrestricted field of fire, but this remained a prototype.

Stronger airframe

In March 1916 AEG flew the 120kW (160hp) Mercedes D.III powered C.IV for the first time. An improved C.II, considerable attention had been spent on production engineering the design, allowing it to be built quickly using semi-skilled labour. Experience with the C.II has demonstrated that greater strength was needed and the C.IV benefitted from a considerably more robust airframe built primarily of metal. In service from 1916, the C.IV was appreciated for its speed and reliability but was not as forgiving an aircraft to fly as the Albatros C.III and DFW C.V, and it never achieved the same popularity as these machines. Nonetheless it performed capably and 70 of the 687 built were still in service at the conclusion of hostilities. A further 609 examples were built of an armoured development known as the J.I, featuring two downward-firing machine guns for the ground attack

AEG C.IV

The AEG was not a particularly elegant machine but was very tough. The majority of the airframe was of welded steel tube covered with fabric. As the Fokker company were also specialists in welded construction and had spare capacity, IdFieg ordered them to build 200 AEG C.IVs, much to the chagrin of Anthony Fokker.

AEG C.IV

Weight: (gross) 1120kg (2469lb)
Dimensions: Length 7.15m (23ft 6in) Wingspan 13.46m (44ft 2in) Height 3.35m (11ft)
Powerplant: One 120kW (160hp) Mercedes D.III 6-cylinder water-cooled inline piston engine
Speed: 158km/h (98mph)
Endurance: 3 hours
Ceiling: 5000m (16,404 ft)
Crew: 2
Armament: One 7.92mm (0.312in) LMG 08/15 'Spandau' fixed firing forward and one 7.92mm Parabellum MG14 machine gun flexibly mounted in rear cockpit; up to 100kg (220lb) bombload

---

role. C.IVs made up a considerable part of the nascent Polish Air Force when 91 were captured during the Greater Polish Uprising of 1918–1919 and saw considerable action in the Polish–Soviet War.

Albatros C.I. to C.III

Albatros produced their first armed aircraft in the form of the C.I that proved highly successful. A developed version, the C.III became the most produced of the Albatros two-seaters.

The C.I was basically an enlarged version of the Albatros B.II designed to utilise one of the more powerful engines that were then becoming available. The prototype flew for the first time during early 1915 with a Benz Bz.III but production aircraft also flew with Mercedes D.III and Argus As.III motors, depending on availability. The increased power available allowed for the fitment of a machine gun for the observer (who was accommodated in the rear cockpit from the start), thus avoiding the cumbersome armament arrangement of contemporary Aviatik aircraft.

The first C.Is arrived at the Front in April of 1915 for reconnaissance, bombing and artillery observation, and proved to be immediately popular, being comparatively fast, structurally robust and easy to fly. The defensive armament was particularly appreciated as interception by Allied scouts, although still comparatively rare in 1915, was becoming more and more commonplace. The C.I seemed an unlikely contender for offensive air-to-air combat, but nonetheless future air aces Oswald Boelcke and Manfred von Richthofen scored their first victories in the type, Boelcke with the aid of his gunner Heinz-Hellmuth von Wühlisch. Richthofen at this time wasn't even a pilot, yet but scored his first victory as an observer in a C.I over the Eastern Front.

Dual machine guns

Remaining in frontline service into early 1917 thanks to its docile handling characteristics, the C.I enjoyed a successful secondary career as a training aircraft and production of the dual-control C.Ib trainer continued at Mercur Flugzeugbau after the aircraft had been taken off operations. By this

Albatros C.III
Weight: (gross) 1353kg (2977lb)
Dimensions: Length 8m (26ft 3in) Wingspan 11.69m (38ft 4in) Height 3.1m (10ft 2in)
Powerplant: One 110kW (150hp) Benz Bz.III or 120kW (160hp) Mercedes D.III or 130kW (180hp) Argus As.III 6-cylinder water-cooled inline piston engine
Speed: 132km/h (82mph)
Endurance: 2.5 hours
Ceiling: 4900m (11,000ft)
Crew: 2
Armament: One 7.92mm (0.312in) LMG 08/15 'Spandau' fixed firing forward and one 7.92mm Parabellum MG14 machine gun flexibly mounted in rear cockpit; up to 90kg (200lb) bombload

Albatros C.III

C.IIIs were in action on the Western Front throughout 1916. This example operated with Staffel 20 of Kampfgeschwader 4 from Carignan airfield in the Ardennes. Operational C.IIIs were usually finished with a varnished fuselage and clear doped wings and tail surfaces.

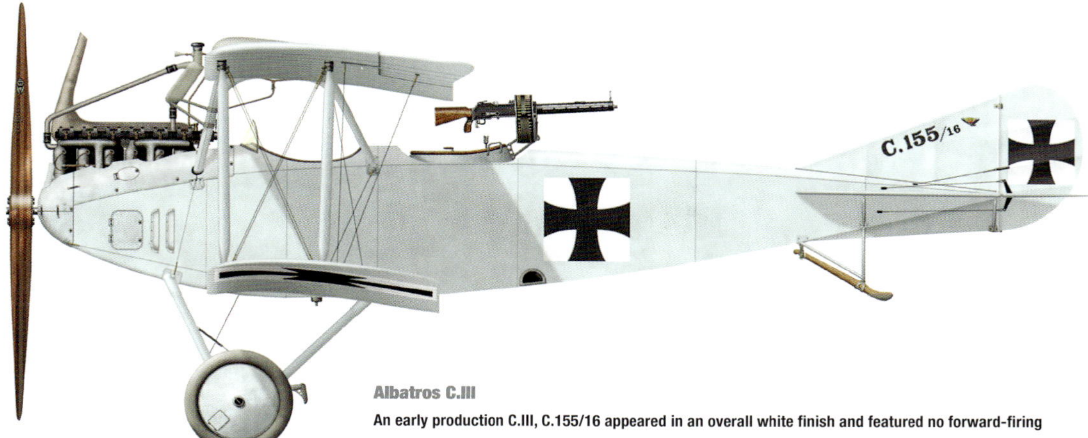

Albatros C.III

An early production C.III, C.155/16 appeared in an overall white finish and featured no forward-firing gun for the pilot. As with many German types, C.IIIs featured a variety of exhaust styles though all were intended to throw the engine exhaust up and away from the crew.

time production had shifted to an improved version, the C.III (the C.II designation being used for a pusher aircraft of which only one example was built), which appeared on the Western Front during December 1915. Easily identified by its elegantly rounded tail surfaces that bestowed greater responsiveness, the C.III introduced a synchronised forward-firing machine gun for the pilot and is believed to be the first German two-seater to feature machine gun armament for both its crew members.

With the same Benz or Mercedes engines as the C.I, it possessed a very similar performance and retained the earlier aircraft's excellent handling and structural strength. Employed in the same roles as its forebear, the C.III grew to command the respect of Allied fighter pilots as it was by no means an easy aircraft to shoot down, due at least in part to its plywood-covered fuselage that could absorb considerably more damage than most of its contemporaries and remain flying.

The absence of a reliable means to aim its bombload with any degree of accuracy meant that its bombing attacks were essentially of nuisance value, but this was always seen as a secondary consideration to its primary

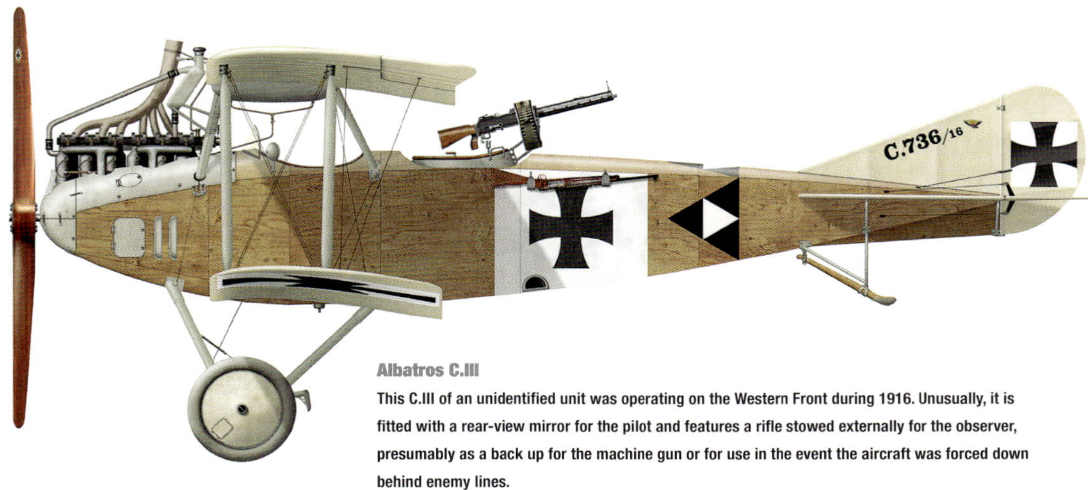

Albatros C.III

This C.III of an unidentified unit was operating on the Western Front during 1916. Unusually, it is fitted with a rear-view mirror for the pilot and features a rifle stowed externally for the observer, presumably as a back up for the machine gun or for use in the event the aircraft was forced down behind enemy lines.

Albatros C.III

Proving that elaborate personal markings were not solely the preserve of the fighter squadrons, this dragon-bedecked C.III operated with Kampfgeschwader 1 in Russia during late 1916. The pilot was Erwin Böhme and the observer was Unteroffizer Lademacher.

reconnaissance role, and the C.III could be equipped with a variety of cameras or wireless equipment for artillery correction. Its usefulness saw very large numbers produced and orders totalling 2271 aircraft are known to have been placed with Albatros and its OAW subsidiary, as well as with five subcontractors. Eventually surpassed by later aircraft in operational units, the C.III had largely disappeared from the Front by mid-1917 but like the C.I before it, production continued for training purposes.

Both types saw use with various air arms after the war, with both Bulgaria and Lithuania constructing copies in 1926–1928 for use as trainers.

Albatros C.III

Weight: (gross) 1353kg (2977lb)

Dimensions: Length 8m (26ft 3in) Wingspan 11.69m (38ft 4in) Height 3.1m (10ft 2in)

Powerplant: One 110kW (150hp) Benz Bz.III or 120kW (160hp) Mercedes D.III or 130kW (180hp) Argus As.III 6-cylinder water-cooled inline piston engine

Speed: 132km/h (82mph)

Endurance: 2.5 hours

Ceiling: 4900m (11,000ft)

Crew: 2

Armament: One 7.92mm (0.312in) LMG 08/15 'Spandau' fixed firing forward and one 7.92mm Parabellum MG14 machine gun flexibly mounted in rear cockpit; up to 90kg (200lb) bombload

Types that could no longer survive in the demanding skies over the Western Front still proved useful in other theatres. This Albatros C.III of Fl.Abt 304 was still providing yeoman service in Palestine during the summer of 1918, a year or so after the type had been withdrawn from combat operations over Europe.

Albatros later C-types

Albatros followed up the success of their C.III with a selection of ever more streamlined two-seaters of greater performance with varying levels of success.

In early 1916, Mercedes developed an eight cylinder development of their six cylinder D.III, the D.IV, which promised 162kW (217hp) and Albatros set about designing an aircraft to utilise it. The new engine was considerably longer, heavier and possessed a different thrust line to the D.III of the Albatros C.III so a completely new aircraft was required rather than the relatively minor modifications that produced the C.III from the C.I.

Albatros C.V

The resulting C.V of 1916 featured a much more streamlined fuselage than the C.III and usefully improved performance. Service use revealed the aircraft was heavy and cumbersome to handle so an improved version was built with balanced ailerons and elevator in addition to the balanced rudder of the original C.V. The aircraft

was also cleaned up aerodynamically with the addition of a wing radiator in place of the 'ear' radiators originally fitted to the fuselage sides. In this form performance was slightly increased and handling was greatly improved. The newer C.V was designated C.V/17 with the original model now referred to as the C.V/16. Unfortunately all was not well with the Mercedes D.IV, which proved unreliable and suffered frequent catastrophic crankshaft failures.

As a result only 424 C.Vs of both types were produced before production switched to the interim C.VII, which married as much of the C.V's excellent airframe to the wholly reliable (although lower powered) Benz Bz.IV. Featuring aspects of the design of both the C.V/16 and C.V/17, and despite being something of a lash-up, the C.VII proved to be a popular and successful aircraft, reportedly possessing

Albatros C.XII

Weight: (gross) 1061kg (2340lb)

Dimensions: Length 8.84m (29ft) Wingspan 14.37m (47ft 2in) Height 3.25m (10ft 8in)

Powerplant: One 190kW (260hp) Mercedes D.III 6-cylinder water-cooled inline piston engine

Speed: 178km/h (110mph)

Endurance: 4 hours 20 minutes

Ceiling: 5640m (18,500ft)

Crew: 2

Armament: One 7.92mm (0.312in) LMG 08/15 'Spandau' fixed firing forward and one 7.92mm Parabellum MG14 machine gun flexibly mounted in rear cockpit

Albatros C.XII

This example of the good-looking but disappointing Albatros C.XII was flown by Hugo Geiger and Theodor Rein of Flieger Abteilung 46b at Marimbois Ferme aerodrome in June 1917. The lightning bolt motif encircling the beautifully streamlined fuselage of the C.XII was Geiger's personal marking.

This official photograph of a factory fresh Albatros C.XII shows well its sleek form and the fine finish of its wooden semi-monocoque fuselage. Unfortunately, though it was well made and streamlined, it proved too weak for sustained operational service.

exemplary handling characteristics and a performance very similar to the higher powered C.V. Less manoeuvrable than the C.III, the C.VII was able to fly higher and outrun Allied fighters on long-range reconnaissance missions and at lower altitudes was quite capable of holding its own in combat with single seaters until the advent of new Allied fighters in the summer of 1917. Around 500 C.VIIs were built, this being the last truly satisfactory Albatros two-seater.

Compromised performance

In mid-1916 the 188kW (217hp) six-cylinder Mercedes D.IVa became available and both Albatros and Rumpler produced aircraft to utilise the new motor. The Albatros C.X consisted of an aerodynamically improved and enlarged C.VII featuring some of the same components as the earlier machine, such as the entire tail unit. During testing it became clear that the Rumpler C.IV was a far superior aircraft to the Albatros, but 400 examples of the C.X were ordered anyway as a back-up to the Rumpler aircraft.

Initially employed for training, IdFlieg had a change of heart and C.Xs were sent to the Front from March of 1917. In service they proved difficult to fly and possessed no real performance advantage over the DFW C.Vs they were supposed to replace. After only a brief period on operations, the C.X was once again relegated to training. The final Albatros two-seater of the war, the C.XII was a refined derivative of the C.X with a lightened structure.

Unfortunately performance, although slightly improved, was still disappointing and the lightened fuselage had a propensity to break in two in the event of a heavy landing. Nonetheless 433 C.XIIs were delivered, although it was withdrawn from the Western Front after only brief service and relegated to training. By contrast, on the less demanding Eastern Front the C.XII served well into 1918.

Aviatik C-types

Aviatik's armed two-seaters were typical of their kind and delivered reliable if unspectacular service until replaced by higher performance machines during 1917.

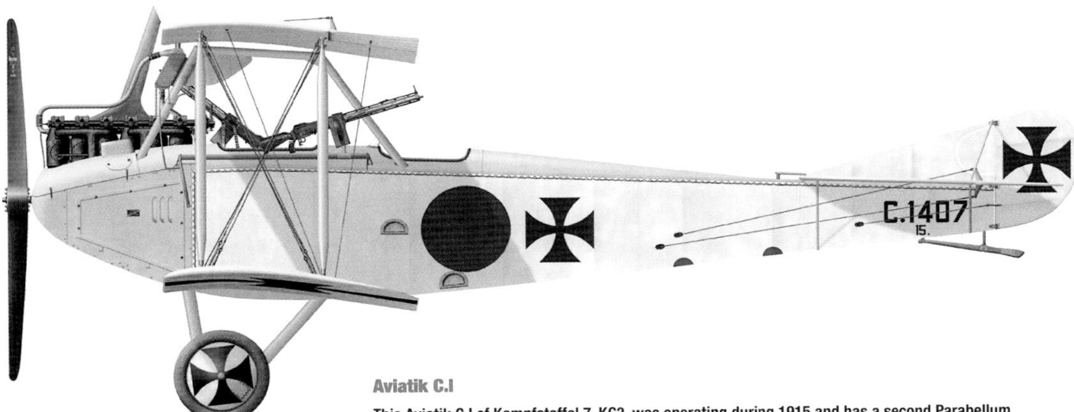

Aviatik C.I

This Aviatik C.I of Kampfstaffel 7, KG2, was operating during 1915 and has a second Parabellum machine gun fitted to the observer's cockpit. Air-to-air combat was relatively rare when the C.I was introduced and the forward-positioned defensive gun was not regarded as a significant flaw.

Aviatik's factory was originally based in Mulhausen (today in France), and in 1914 perilously close to the front line, so in the autumn of that year Aviatik evacuated their operations to Freiburg. Once there, the company flew its first armed two-seater, the C.I, in April 1915. A totally conventional aircraft, the C.I was arranged, like the equivalent British BE.2, with the observer in the front cockpit and pilot at the rear. The flexibly mounted Parabellum machine gun was clipped to a sliding rail on either side of the cockpit that featured a quick-release mechanism allowing the observer to transfer the gun to whichever side it was required. This system was unwieldy and the gun's position resulted in a badly restricted field of fire. The observer's view was also restricted, particularly downwards, as his position was directly above the lower wing, a serious flaw for an aircraft intended primarily to observe operations on the ground.

Design iterations

The C.I was produced in large numbers, 402 by Aviatik and a further 146 by Halberstadt under licence. An aerodynamically improved version, the Benz B.III powered C.II was built in small numbers before production switched to the definitive C.III. This featured the same Mercedes D.III as the C.I in a new, more streamlined cowling and with a large spinner fitted over the propeller boss. The radiator was now an aerofoil-shaped unit in the starboard upper wing and the exhaust discharged horizontally to starboard, both these changes improving forward view considerably. The improved streamlining imparted a useful increase in performance and the C.III began to reach the Front during 1916. Despite its obvious

Aviatik C.I

Weight: (gross) 1340kg (2954lb)

Dimensions: Length 7.93m (26ft) Wingspan 12.5m (41ft) Height 2.95m (9ft 8in)

Powerplant: One 119kW (160hp) Mercedes D.III 6-cylinder water-cooled inline piston engine

Speed: 142km/h (88mph)

Endurance: 3 hours

Ceiling: 3500m (11,500ft)

Crew: 2

Armament: One 7.92mm (0.312in) Parabellum MG14 machine gun flexibly mounted in forward cockpit

disadvantages, the observer's cockpit remained in the forward position. Late production C.IIIs eventually moved the observer and his defensive machine gun to the rear cockpit but by then other superior aircraft were in production. Aviatik were engaged in building the excellent DFW C.V from 1916, and the Aviatik C-types were withdrawn from operations during 1917.

DFW C-types

Built in greater numbers than any other German type, the DFW C.V was an extremely effective two-seater and enjoyed a particularly long service career for an aircraft of its era.

Deutsche Flugzeug-Werke (DFW) was founded in 1910, initially building licensed Farman designs and later becoming one of the many German companies to produce Taubes before building its own original designs. It is no longer known exactly when the DFW C.I was first flown but this event likely took place in early 1916. The C.I was a conventional general purpose type featuring the then-standard arrangement of pilot in the rear and observer/gunner in the front cockpit. These positions were sensibly reversed in the otherwise identical C.II. The C.III is believed to be the designation applied to an experimental pusher design but the C.IV was a developed C.II featuring a single-bay wing design. Production of all the early DFW C-types was modest, as similar machines by other manufacturers, particularly Albatros and Rumpler, were considered superior, but an increase in engine power and improved airframe resulted in the highly successful C.V.

The C.V reverted to a two bay biplane wing and featured a 149kW (200hp) Benz Bz.IV delivering a decent overall performance but with particularly good climb and handling for a 1916 two-seater. The fuselage sides and underside were constructed of plywood

DFW C.V

Weight: (gross) 1430kg (3153lb)

Dimensions: Length 7.88m (25ft 10in) Wingspan 13.27m (43ft 6in) Height 3.25m (10ft 8in)

Powerplant: One 150kW (200hp) Benz Bz.IV 6-cylinder water-cooled inline piston engine

Speed: 155km/h (96mph)

Range: 300km (186 miles)

Endurance: 3.5 hours

Ceiling: 5000m (16,400ft)

Crew: 2

Armament: One 7.92mm (0.312in) LMG 08/15 'Spandau' fixed firing forward and one 7.92mm Parabellum MG14 machine gun flexibly mounted in rear cockpit; up to 100kg (220lb) bombload

sheets with the smooth curved upper surface formed from mouldings of strip ply, then covered with doped fabric proving light but very sturdy. The wings were of conventional wood, wire and fabric construction and the tail featured a steel tube structure.

Ordered into production in August 1916, the first C.Vs reached the Front in late September 1916 and although the C.V was intended as a reconnaissance and observation platform, it proved exceptionally versatile and over the next two years would be used for bombing, as a fighter and ground attack aircraft and as a trainer. For its original role it could be fitted with a variety of cameras and wireless equipment and later aircraft benefited from an internal bomb rack

and an aerodynamically improved nose with a streamlined spinner on the propeller hub.

Approximately 3,900 C.Vs were built, most by DFW but licence production by Aviatik, Halberstadt, LVG and Schütte-Lanz added many more. Aviatik alone are known to have supplied 1400 examples over the course of 12 production orders.

Formidable enemy

The airframe was notably robust and the C.V was blessed with excellent manoeuvrability, sufficient to allow it to outmanoeuvre most Allied fighters. Indeed, in the hands of an experienced crew the C.V was a difficult opponent for even the best fighter pilots, for example James McCudden, the seventh most successful Great War fighter pilot of any nation, described an inconclusive combat with a C.V in his SE.5a that took place on 12 December 1917: "I fought that D.F.W. from 4000 feet down to 500 feet over our lines, and at last I broke off the combat, for the Hun was too good for me and had shot me about a lot. Had I persisted he certainly would have got me." That an airman of McCudden's abilities should have met his match in the form of a seemingly relatively innocuous reconnaissance machine speaks volumes about the quality of this exceptionally useful aircraft and its crew.

Front-line service

Although comparatively old by November 1918 and gradually being replaced by newer aircraft, around 600 C.Vs were still in front-line service in the West when the war came to an end. By this time increasing numbers were used for training and a dedicated variant was produced with the less powerful 150kW (200hp) C.III NAG engine known as the DFW C.Vc. DFW

DFW C.V

Believed to have been operated by Flieger-Abteilung (Artillery) 282 in 1917 or early 1918, this mid-production C.V displays the original *eisernkreuz* national markings. C.Vs flew countless missions spotting for artillery units during the last two and a half years of the war.

were working on the improved C.VI in late 1918 but only one appears to have been built before the end of hostilities. In addition to its use by Germany, two were flown by White forces during the bitter Civil war of 1918 in Finland and a handful were operated by Bulgaria.

Subsequently seven unlicensed copies called the DAR Uzanov-1 were built by the Bulgarian State Workshops as late as 1925 for use as trainers. Six further air forces operated C.Vs after 1918, notably Poland, which captured or bought some 63 examples, many of which saw combat during the Polish–Soviet war of 1919–1921.

Three further developed airframes were constructed at DFW, designated F.37 by DFW, and during 1919 one of these was fitted with a 220kW (295hp) BMW IV engine and used to set an altitude record of 7700m (25,260ft). Unfortunately this could not be officially recognised as it was in breach of the terms of the armistice. The same aircraft was later converted into a three passenger civil aircraft with a richly upholstered enclosed cabin called the DFW P.1 Limousine, but no production ensued.

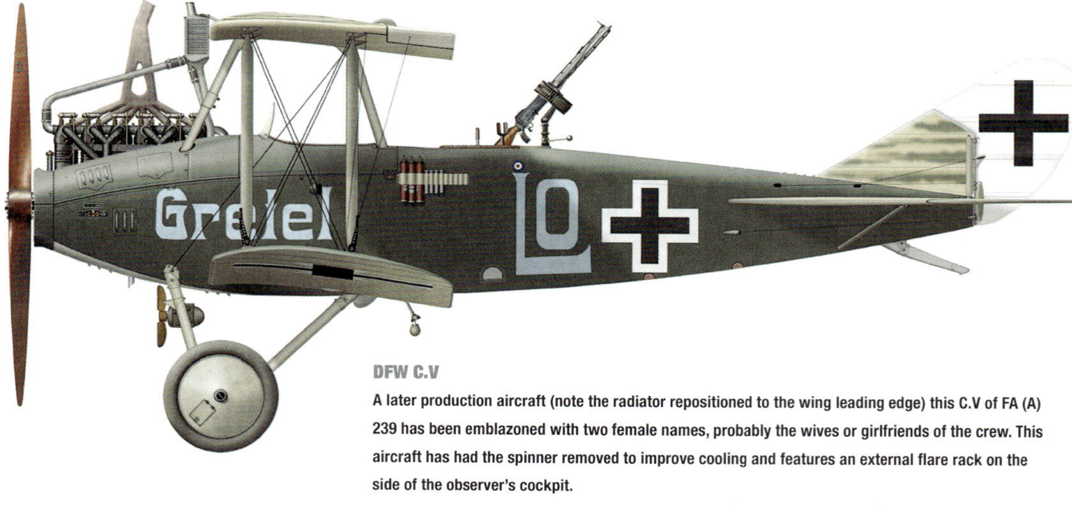

DFW C.V

A later production aircraft (note the radiator repositioned to the wing leading edge) this C.V of FA (A) 239 has been emblazoned with two female names, probably the wives or girlfriends of the crew. This aircraft has had the spinner removed to improve cooling and features an external flare rack on the side of the observer's cockpit.

DFW C.V

Another late production example, this C.V was built by Aviatik and operated by Flieger-Abteilung (Artillery) 223 in May 1918. This aircraft was captured by British forces and subsequently flown with RAF markings, although its new owners preserved the individual butterfly marking on the fuselage.

Halberstadt C.V

One of the last C-types to enter service in numbers, the Halberstadt C.V was probably Germany's finest general purpose two-seater to see service during the war.

Derived from the C.IV that did not enter production, production of Halberstadt's rugged C.V resulted from implementation of the 'America Programme' of 1918, which sought to rapidly expand Germany's aviation units as the USA entered the war. Intended for long-range, high-altitude reconnaissance work, the C.V resembled Halberstadt's CL series but was considerably larger. The prototype was tested over March and April 1918 at Adlershof before being ordered into production, with the first aircraft arriving at front-line units during June. Its excellent performance resulted in further orders and the type was built under licence by Aviatik, BFW and DFW. A total of around 550 was constructed by the armistice, a creditable amount given that the aircraft was in production for less than five months.

High-altitude flight

A large aircraft, the C.V featured a high compression Benz Bz.IVü and long span wings, roughly twice as long as the fuselage, to provide good performance and handling at altitude. Unusually for an aircraft of its era, some consideration was made for the comfort of the crew at the heights it was expected to operate and the C.V boasted an electrically heated cockpit. A variety of cameras could be fitted; a laterally sliding door in the observer's cockpit floor exposed the lens when the camera was in use. Flying at 5000m (16,404ft) or higher, the C.V was extremely difficult to intercept by Allied fighters and the type proved highly popular with its crews. A developed version, the C.VIII, with a 180kW (174hp) Maybach engine and a staggering 9000m (29,530ft) service ceiling was poised to enter production

at the armistice but remained a prototype. After 1918, at least seven countries operated captured or interned examples of the C.V, the Soviet Union purchasing 18 directly from Halberstadt as late as 1922.

Halberstadt C.V

Weight: (gross) 1635kg (3605lb)

Dimensions: Length 6.92m (22ft 8in) Wingspan 13.62m (44ft 8in) Height 3.36m (11ft)

Powerplant: One 160kW (220hp) Benz Bz.IVü 6-cylinder water-cooled inline piston engine Speed: 170km/h (110mph)

Endurance: 3 hours 30 minutes

Ceiling: 6500m (21,325ft)

Crew: 2

Armament: One 7.92mm (0.312in) LMG 08/15 'Spandau' fixed firing forward and one 7.92mm Parabellum MG14 machine gun flexibly mounted in rear cockpit; up to 50kg (110lb) bombload

---

Halberstadt C.V

Despite seeing production and service in considerable numbers, the service use of the Halberstadt C.V remains little known. This is due, at least in part, to the chaotic conditions of the German retreat in which the C.V spent most of its service life. This C.V served with Flieger Abteilung 46b during 1918.

LFG Roland C.II & C.IV

The Roland C.II and C.IV were aerodynamically ahead of their time and boasted an unusually good performance but at the cost of somewhat capricious handling characteristics.

LFG (Luftfahrzeug-Gesellschaft) was founded in 1908 as an airship construction works, branching into aircraft construction in 1913 and producing new designs under the trade name Roland to avoid confusion with LVG (Luftverkehrsgesellschaft). Their first and most successful design was the C.II in which great care had been taken to streamline the aircraft to as great a degree as possible.

Wire bracing was kept to a minimum and the aircraft featured broad cord 'I' struts between the wings that caused less drag than conventional struts. There were even modest fillets, fairing the wings to the fuselage, a feature that smooths turbulent boundary layer airflow, reducing drag and wake turbulence, and one that would be absent from many aircraft designs for the next two decades.

Field of fire

The C.II featured LFG's patented *wickelrumpf* fuselage that was created in halves from two layers of thin plywood strips, each layer applied at an opposing angle of around 60 degrees formed over a mould. Each half of the fuselage was glued and tacked onto the internal framework, the centreline seams were taped over and then the whole fuselage was covered with doped-on fabric. This construction method allowed for complex compound curves to be formed and imparted great strength for light weight but it was time consuming and expensive to build. The C.II featured a large but streamlined fuselage that completely filled the gap between the upper and lower wings, obviating the need for cabane struts. The crew positions gave them a commanding view upwards and sideways with an excellent field of fire for the observer but visibility downwards was poor, ameliorated somewhat by large windows cut into the fuselage sides. The C.II flew in October 1915, with the first batch of 50 of an eventual 400 ordered in December.

LFG Roland C.II

Weight: (gross) 1284kg (2831lb)
Dimensions: Length 7.7m (25ft 3in) Wingspan 10.3m (33ft 10in) Height 2.9m (9ft 6in)
Powerplant: One 120kW (160hp) Mercedes D.III 6-cylinder water-cooled inline piston engine
Speed: 165km/h (103mph)
Endurance: 4 hours
Ceiling: 4000m (13,000ft)
Crew: 2
Armament: One 7.92mm (0.312in) Parabellum MG14 machine gun flexibly mounted in rear cockpit; up to 50kg (110lb) bombload; later production aircraft added one 7.92mm LMG 08/15 'Spandau' fixed firing forward

---

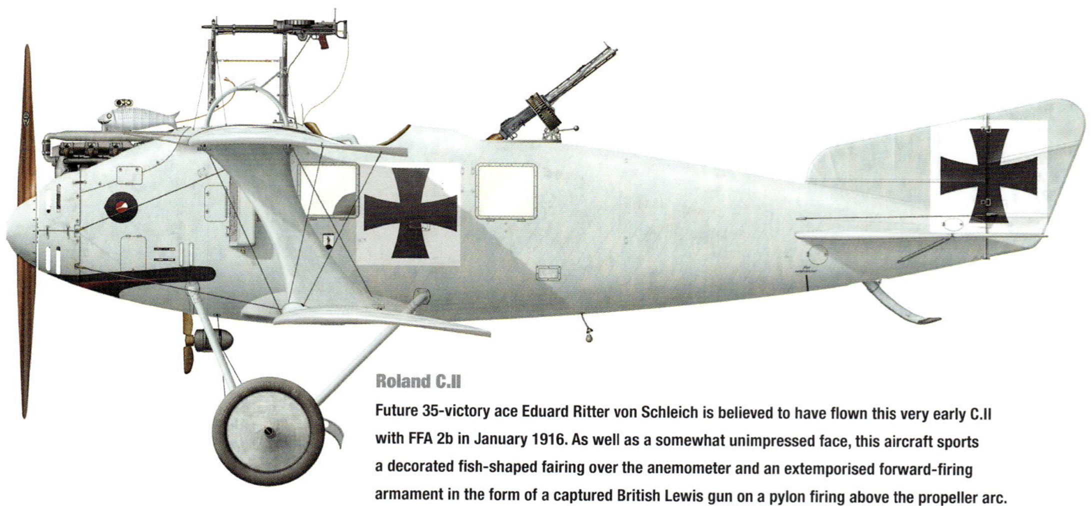

Roland C.II

Future 35-victory ace Eduard Ritter von Schleich is believed to have flown this very early C.II with FFA 2b in January 1916. As well as a somewhat unimpressed face, this aircraft sports a decorated fish-shaped fairing over the anemometer and an extemporised forward-firing armament in the form of a captured British Lewis gun on a pylon firing above the propeller arc.

'The Whale'

Nicknamed Walfisch (whale) due to its corpulent yet streamlined shape, the C.II was as fast as it looked, boasting a top speed some 30km/h faster than contemporary two-seaters and faster than many fighters. When the C.II arrived at the Front in March 1916 it caused something of a sensation thanks to its striking appearance and the fact that it was the smallest and fastest two-seater in the German armoury. Speed performance was roughly equivalent to the opposing Sopwith Pup and Nieuport 17, though the larger Roland was considerably less manoeuvrable then either Allied single-seater. The type garnered the respect of its adversaries – British ace Albert Ball, who managed to down a C.II for the first of his 44 victories, described it as, "the best German machine now" during 1916.

Unfortunately for its crew, the large fuselage restricted airflow over the tailplane that affected manoeuvrability and their position virtually atop the wing restricted visibility forwards and downwards. Pilot view of the airfield on the approach to landing

Roland C.IIa

Not until the advent of the Curtiss P-40 would an aircraft be so regularly decorated with a shark's mouth as the C.II. This later C.IIa was operating with Schusta 6 in early 1917. As well as its toothy grin, the aircraft sports some dainty curtains painted onto the celluloid side windows. The Walfisch seems to have inspired more irreverent decoration than any other Great War aircraft.

was virtually non-existent and led to a disproportionate number of accidents.

On the second production batch a synchronized machine gun was introduced as an offensive weapon for the pilot, this necessitating a redesign of the roll-over pylon above the cockpit. Alterations were also made to the wing, shortening the span and moving the interplane struts inboard to strengthen the wing, resulting in the aircraft being redesignated Roland C.IIa. The final batch of C.IIas also featured an enlarged tail to improve control response, which was a long-overdue change.

By the time these late production C.IIas reached the Front in early 1917 the Walfisch had lost its performance advantage and the type would be gradually withdrawn to training schools over the course of the year, serving in this role into 1918.

LVG C.II

The first German aircraft to be armed with a machine gun for the observer, the LVG C.I and its near identical derivative, the C.II, gave yeoman service as a reconnaissance and light bombing machine.

Luftverkehrsgesellschaft had been founded in 1912 to build Farman aircraft under licence, subsequently producing the original designs of their Swiss chief designer, Franz Schneider. During September 1914 Schneider developed a machine gun ring mounting, allowing the weapon to be fired in any direction above the fuselage and downwards to either side. This was mounted on a strengthened LVG B.I airframe and resulted in it becoming the first C-type armed aircraft. Eventually, Schneider's ring mount would become a standard feature of all German C-types.

Unreliable engine

The C.I was in other regards a very typical German two-seater for its era. Limited production ensued before switching to the much more common C.II that appeared towards the end of 1915 and was built in numbers. Production totals are unknown but the aircraft was built under licence by both Ago and Otto, suggesting that several hundred were built, and it is known that 250 examples of the C.I and II were operational in the spring of 1916, making it one of the most numerous aircraft in service. The C.III inexplicably featured the positions of the pilot and observer reversed and did not enter production, but the slightly enlarged eight-cylinder Mercedes D.IV powered LVG C.IV did, earning its place in aviation history by becoming the first fixed-wing aircraft to bomb London, when six bombs were dropped near Victoria Station on 28 November 1916. The unfortunate crew had little time to savour their achievement as the aircraft was damaged by a French anti-aircraft battery on its return journey (although some sources suggest that it suffered engine failure) and the crew were taken prisoner following the forced landing. The unreliability of the Mercedes engine meant that very few C.IVs were built.

LVG C.II

Weight: (gross) 1405kg (3097lb)

Dimensions: Length 8.10m (26ft 7in) Wingspan 12.85m (42ft 2in) Height 2.93m (9ft 7in)

Powerplant: One 120kW (160hp) Mercedes D.III 6-cylinder water-cooled inline piston engine

Speed: 130km/h (81mph)

Endurance: 4 hours

Ceiling: 4000m (13,125ft)

Crew: 2

Armament: One 7.92mm (0.312in) Parabellum MG14 machine gun flexibly mounted in rear cockpit; up to 60kg (132lb) bombload; later production aircraft added one 7.92mm LMG 08/15 'Spandau' fixed firing forward

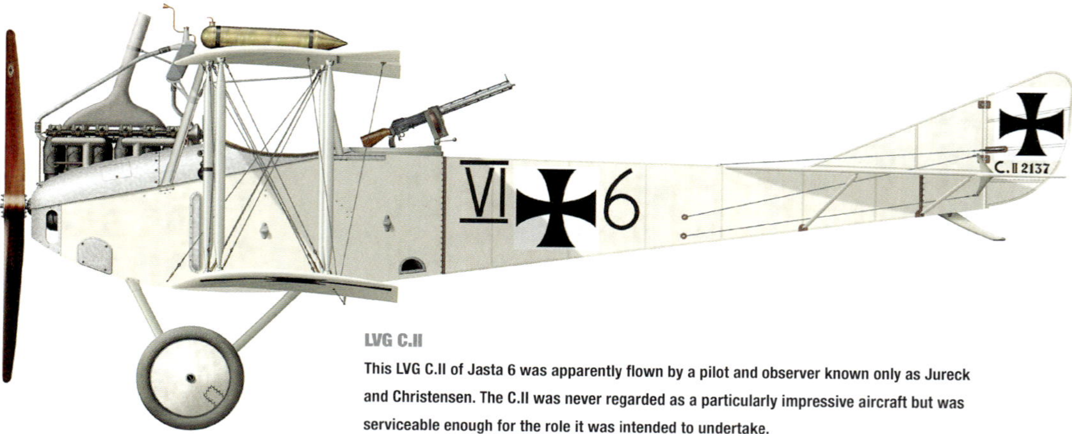

LVG C.II

This LVG C.II of Jasta 6 was apparently flown by a pilot and observer known only as Jureck and Christensen. The C.II was never regarded as a particularly impressive aircraft but was serviceable enough for the role it was intended to undertake.

LVG C.V & C.VI

The LVG C.V and C.VI complemented the high-flying Rumplers, fulfilling the need for a sturdy and reliable aircraft for light bombing, artillery observation and medium-range reconnaissance.

LVG C.VI

In unusual winter camouflage applied over the varnished wood fuselage skin, this C.VI was flown by Leutnant Kuchenthal and Unteroffizer Friedmeyer in the last weeks of the conflict. Note the holstered flare pistol and externally mounted compass to the rear of the observer's cockpit.

Developed from the C.II, the LVG C.V was an order of magnitude more formidable, powered by the 150kW (200hp) Benz Bz.IV and first taking to the air in early 1917. Entering service in the summer of 1917, the C.V was popular with crews due to its good handling qualities and robust construction, although visibility from the cockpit was poor in some directions, especially forwards.

A large aircraft, the C.V was nonetheless able to give a good account of itself in aerial combat. Production totals are now unknown but several hundred are believed to have been built.

Improved performance

In February 1918 LVG flew the first example of a developed version of the C.V, the C.VI. Powered by the same Benz Bz.IV engine, the C.VI featured a lightened and more compact fuselage for a modest improvement in performance and the aircraft entered service at the Front in June. Despite this relatively late start, mass production built up quickly and 1100 examples of the C.VI were built before the armistice. Like the C.V before it, the C.VI was highly regarded by its crews for its respectable climb rate, decent speed and good manoeuvrability.

As with many late war German types, both the C.V and C.VI continued to serve in the air arms of various nations, particularly in Eastern Europe and the Baltic states. Some 150 C.Vs were captured by Poland, many of which were used in combat with the Soviet Union, whereas Lithuanian examples of the C.VI were maintained

LVG C.VI

Weight: (gross) 1390kg (3064lb)
Dimensions: Length 7.45m (24ft 5in) Wingspan 13m (42ft 8in) Height 2.85m (9ft 4in)
Powerplant: One 150kW (200hp) Benz Bz.IV 6-cylinder water-cooled inline piston engine
Speed: 170km/h (110mph)
Endurance: 3 hours 30 minutes
Ceiling: 6500m (21,300ft)
Crew: 2
Armament: One 7.92mm (0.312in) LMG 08/15 'Spandau' fixed firing forward and one 7.92mm Parabellum MG14 machine gun flexibly mounted in rear cockpit; up to 90kg (200lb) bombload

in service until 1940, by then quite likely the oldest aircraft designs still in military service, a testament to the qualities of the LVG design.

An airworthy C.VI was restored to fly in the UK in 1937 and was maintained in flying condition from 1972 until its final flight before being permanently grounded in June 2003. It was the last genuine World War I German aircraft in airworthy condition.

LVG C.VI

'Lotta' was operated by FFA 233 and bears this unit's distinctive nose markings. Note the wind-driven generator on the undercarriage leg, electricity being required both for radio equipment and the crew's heated flying suits, and the box for carrying a homing pigeon strung between the interplane struts.

Many LVG C.VIs enjoyed a second career as civil aircraft carrying passengers and mail. Some were converted with an enclosed cabin for passengers but this example remains essentially unchanged from its wartime configuration (though undoubtedly much cleaner) requiring its passengers to don warm clothing, like these two hardy souls.

Rumpler C.I

The Rumpler C.I was the fastest of the armed two-seaters that began to appear during 1915. Popular due to its excellent performance and sturdy construction, the Rumpler enjoyed a long and effective career.

Rumpler C.I

Arguably the best of the crop of C-type two-seaters that appeared during 1915, and a reasonable contender for best general purpose reconnaissance aircraft in the world at the time, this Rumpler C.I served with Kagohl 4 during the cataclysmic Battle of Verdun in the spring of 1916.

Notable for being faster than the contemporary Fokker Eindecker fighter and possessing a similar rate of climb, the C.I first took to the air during the summer of 1915. Acceptance testing was completed by the end of October, the aircraft having already been ordered into production in July, and the first C.Is started to arrive at the Front during November and December 1915. As was the case with its contemporaries, the C.I flew with both 120kW (160hp) Mercedes D.III and 112kW (150hp) Benz B.III engines. Additionally a 134kW (180hp) Argus As.III-powered version was also built under licence by Hannover, which offered better low-altitude performance at the expense of reliability at altitude.

Robust design

On operational service the type was well-liked by crews, its robust and damage-resistant steel tube and plywood fuselage conferred great survivability in the event of a crash, while at the same time the aircraft offered better performance and easier handling than its contemporaries. In service throughout 1916, the advent of higher-performance aircraft saw it gradually reassigned from the Western Front to the training role though its continued combat usefulness in less demanding areas is demonstrated by the fact that new C.Is were being supplied to the Front in the Middle East as late as the spring of 1917, where they proved highly effective.

The C.I was also the basis for a modestly successful seaplane fighter design, the Rumpler 6B, of which 88 were built. Modified with forward stagger to the wings, the observer's cockpit removed and with a larger rudder to offset the increased side area caused by the addition of floats, the aircraft were mostly based at Ostend and Zeebrugge, although some operated in the Black Sea against the Russians. One 6B was used by Finland until 1925.

Rumpler C.I

Weight: 1333kg (2939lb)

Dimensions: Length 7.85m (25ft 9in) Wingspan 12.15m (39ft 10in) Height 3.06m (10ft)

Powerplant: One 120kW (160hp) Mercedes D.III 6-cylinder water-cooled inline piston engine

Speed: 152km/h (94mph)

Endurance: 4 hours

Ceiling: 5050m (16,570ft)

Crew: 2

Armament: One 7.92mm (0.312in) Parabellum MG14 machine gun flexibly mounted in rear cockpit; up to 100kg (220lb) bombload; later production aircraft added one 7.92mm LMG 08/15 'Spandau' fixed firing forward

Rumpler C.III & C.IV

Possessing exceptional altitude performance, the Rumpler C.IV was arguably the finest reconnaissance aircraft of the war and virtually immune from fighter interception for much of its service life.

Rumpler expended a great deal of effort in perfecting the successor to its highly popular C.I and this work initially saw the light of day in the form of the C.III, a Benz Bz.IV powered aircraft with a new wing planform and aerofoil section optimised for high-altitude performance. The aircraft was a qualified success with performance showing an improvement over the C.I, although only 75 or so were built and it is essentially regarded as a developmental stepping stone to the outstanding C.IV.

Reconnaissance asset

The C.IV utilised the new Mercedes D.IVa engine combined with a modified late-production C.III airframe, the most obvious change being the wingtips of the lower wing that were altered to a rounded shape to lessen induced drag. The result was an aircraft of unmatched altitude performance that would become Germany's most important strategic reconnaissance asset for the latter part of the war. First flown in late September 1916, the prototype C.IV dramatically demonstrated its potential in October by climbing to 5000m (16,404ft) in 33 minutes with a full combat load and maintained 165km/h (102mph) at altitude, an outstanding performance for the time. IdFlieg type testing was complete by the end of January 1917 and the C.IV was approved for production on 10 February, the first aircraft reaching the Front later the same month.

Early service confirmed the Rumpler's excellent performance but led to concerns about the light structure, as the aircraft suffered fuselage fractures and the airframe was prone to twisting and flexing during

Rumpler C.III

Weight: (gross) 1264kg (2780lb)
Dimensions: Length 8.20m (26ft 11in) Wingspan 12.66m (41ft 6in) Height 3.25m (10ft 8in)
Powerplant: One 150kW (200hp) Benz Bz.IV 6-cylinder water-cooled inline piston engine
Speed: 136km/h (85mph)
Endurance: 3 hours 30 minutes
Ceiling: 4000m (13,100ft)
Crew: 2
Armament: One 7.92mm (0.312in) LMG 08/15 'Spandau' fixed firing forward and one 7.92mm Parabellum MG14 machine gun flexibly mounted in rear cockpit; up to 100kg (220lb) bombload

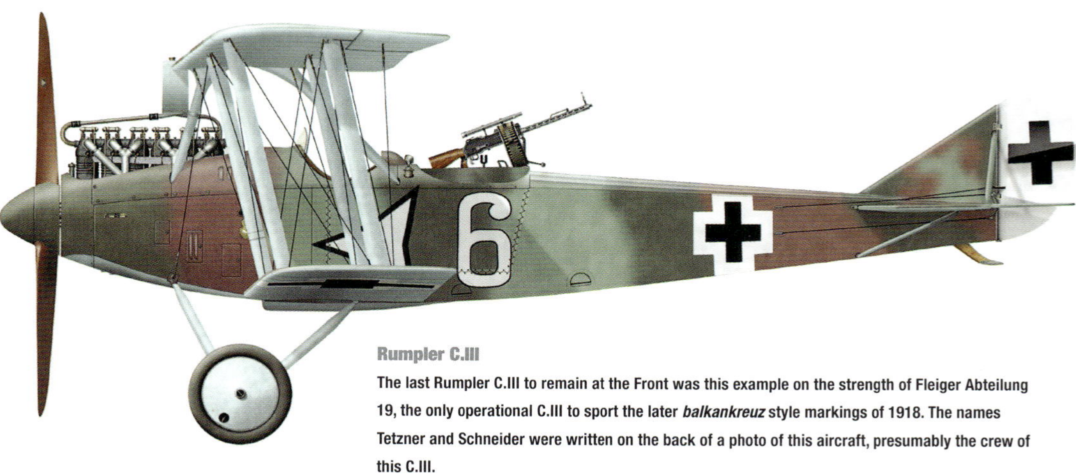

Rumpler C.III

The last Rumpler C.III to remain at the Front was this example on the strength of Fleiger Abteilung 19, the only operational C.III to sport the later *balkankreuz* style markings of 1918. The names Tetzner and Schneider were written on the back of a photo of this aircraft, presumably the crew of this C.III.

Rumpler C.IV

The exhortation 'Good People don't shoot' appeared on several German aircraft both in German and English (one wonders if it had the desired effect?). This C.IV is a later production example from a February 1917 order with bluntly curved nose and no spinner and features lozenge fabric over most of the fuselage and wings.

manoeuvres. During April all C.IVs were returned to the factory to receive plywood fuselage reinforcements. The aircraft was also found to possess a slow aileron response and was prone to stall abruptly, with recovery from the ensuing spin proving difficult. The handling issues were generally considered to be outweighed by the Rumpler C.IV's excellent performance, although a batch of C.IVs built by Pfalz intended primarily for the training role featured ailerons on all four wingtips rather than solely on the top wing to improve aileron responsiveness. This improvement was never applied to C.IVs built by other manufacturers however.

Impressive altitude

Operational C.IVs routinely flew at around 5000m (16,404ft) or more, a height most Allied fighters were incapable of reaching. The few that could were generally outrun or outmanoeuvred by the Rumplers at this altitude. With the C.IV enjoying considerable success at the Front, development continued to improve its performance still further. The more powerful Basse & Selve BuS.IVa became available in late 1917 and was fitted to the C.IV when supplies allowed. More dramatically, some C.IVs were fitted with the 'overcompressed' Maybach Mb.IVa that, although rated as a lower-powered engine than the standard Mercedes at sea level, actually improved altitude performance due to its ability to maintain power at height. The Maybach engine was also used for the developed C.VII that featured a slightly smaller fuselage and wings of slightly greater area. Flying at 7000m (22,966ft) the C.VII featured a generator to supply power for electrically heated flight suits and oxygen equipment for the crew. Stripped of the forward gun and all extraneous equipment, a modified C.VII named the Rubild (a contraction of Rumpler Bildaufklärer) was capable of operating at 7300m (23,950ft) and utilised the Messter strip camera to map vast areas, including photographing the entire Western Front every two weeks by the end of the war. Around 1000 C.IVs, C.VIIs and Rubilds were built.

Rumpler C.IV

Weight: (gross) 1530kg (3373lb)
Dimensions: Length 8.41m (27ft 7in) Wingspan 12.66m (41ft 6in) Height 3.25m (10ft 8in)
Powerplant: One 190kW (260hp) Mercedes D.IVa 6-cylinder water-cooled inline piston engine
Speed: 171 km/h (106mph)
Endurance: 3 hours 30 minutes
Ceiling: 6400m (21,000ft)
Crew: 2
Armament: One 7.92mm (0.312in) LMG 08/15 'Spandau' fixed firing forward and one 7.92mm Parabellum MG14 machine gun flexibly mounted in rear cockpit; up to 100kg (220lb) bombload

GROUND ATTACK & ESCORT FIGHTERS

Germany pioneered the use of aircraft as a weapon to directly attack ground forces. The CL-types, which excelled in the ground-attack role despite their intended use as escort fighters, were augmented by the tough armoured J-types that paved the way for the kind of close air support perfected for the *Blitzkrieg* by the Ju 87 Stuka some 20 years later.

This chapter includes the following aircraft:

- Halberstadt CL.II
- Halberstadt CL.IV
- Hannover CL-types
- Junkers CL.I
- Albatros J.I & J.II
- Junkers J.I

Displaying its unusual biplane tail, adopted to maximize the observer's field of fire to the rear, the Hannover CL.III was one of the finest close-support types to see action during the war. German application of tactical air power against ground targets was arguably more sophisticated than that of the Allies and proved extremely effective.

Halberstadt CL.II

Intended originally to operate as an escort fighter, a role in which it excelled, Halberstadt's versatile CL.II aircraft also proved extremely successful in the ground-attack role.

In 1916 Halberstädter Flugzeugwerke designed a two-seat escort fighter or 'leicht' (lightweight) C-type, IdFlieg's newly created CL-type. After passing official tests in May 1917, production CL-IIs started to reach the Front in August and proved immediately successful, with excellent manoeuvrability, good climb rate and a wide field of fire for the rear gunner, enabling them to engage enemy single-seaters on roughly equal terms. Unusually, the crew shared one large cockpit, this feature allowing for easy communication between the two and contributed to its popularity amongst crews. The CL-IIs initially formed the mainstay of the Schutzstaffeln, escorting C-type reconnaissance aircraft, but the close-support fighter role assumed greater importance, and in March 1918 CL-II units were renamed as Schlachtstaffeln reflecting the shift to infantry support.

Ground attack role

The elevated gun ring allowed the observer to engage ground targets and a rack was fitted to the port side of the fuselage with space for 10 stick grenades. The CL-II was highly effective as soon as it was committed, in September 1917 for example, 24 Schlachtstaffeln Halberstadts attacked British troops advancing over the Somme bridges at Bray and St. Christ

Halberstadt CL.II

Weight (gross): 1133kg (2498lb)
Dimensions: Length 7.3m (23ft 11in) Wingspan 10.77m (35ft 4in) Height 2.75m (9ft)
Powerplant: One 120kW (160hp) Mercedes D.III 6-cylinder water-cooled inline piston engine Speed: 165km/h (103mph)
Endurance: 3 hours
Ceiling: 5090m (16,700ft)
Crew: 2
Armament: One 7.92mm (0.312in) LMG 08/15 'Spandau' fixed firing forward and one 7.92mm (0.312in) Parabellum MG14 machine gun flexibly mounted in rear cockpit; 10 stick grenades; up to five 10kg (22lb) Wurfgranaten 15 trench mortar fragmentation bombs

Halberstadt CL.II

As flamboyant as its fighter contemporaries in the Jagstaffeln, this impressive CL.II of Schlasta 21 was being flown by Rudolf Kolodzicj and Max Niemann when they were shot down and captured on 2 October 1918. The aircraft bears the name 'Else' on the starboard side of the nose and 'Martha' to port.

Halberstadt CL.II

Observer Fridolin Redenbach 'commanded' CL.II 'Anni' of Schutzstaffel 27b in late 1917, observers often being the ranking officer of a two-seater crew. The aircraft features Halberstadt's unusual 'stipple' camouflage over a painted lozenge pattern on the fuselage. The significance, if any, of the terrifying beast remains unknown.

causing panic. Troops jumped over the parapets of the two bridges to escape the machine-gun fire and grenades from the enemy aircraft. The artillery troops and their horses in the rear were also attacked and it was estimated the CL.IIs had disrupted the best part of a division. The Halberstadts were attacked by two Sopwith Scouts, one of which was shot down for no losses to the CL.IIs.

Halberstadt built 700 CL-IIs before switching to the CL-IV in mid 1918. A further 200 CL.IIs were subsequently built by BFW, the aircraft being used until the end of the war.

The lack of any unit markings suggest this Halberstadt CL.II is a factory fresh example, not yet assigned to the front. This view shows well the unusual elevated ring mounting for the observer's machine gun over the rear part of the cockpit opening, which allowed the flexibly-mounted Parabellum to fire directly downwards on either side of the aircraft.

Halberstadt CL.IV

Developed from the CL.II, the CL.IV was optimized for the ground attack role and, like its forebear, proved extremely effective.

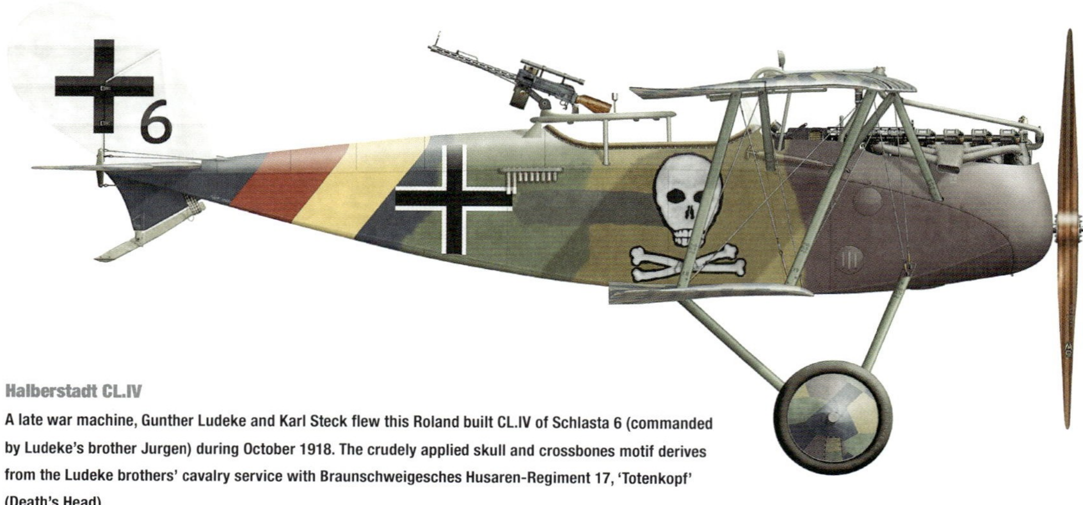

Halberstadt CL.IV

A late war machine, Gunther Ludeke and Karl Steck flew this Roland built CL.IV of Schlasta 6 (commanded by Ludeke's brother Jurgen) during October 1918. The crudely applied skull and crossbones motif derives from the Ludeke brothers' cavalry service with Braunschweigesches Husaren-Regiment 17, 'Totenkopf' (Death's Head).

Intended to supplement and eventually replace the CL.II, the CL.IV possessed superior manoeuvrability, essential for an unarmoured aircraft that relied on agility as its primary means of defence. It featured a more compact, lightened fuselage and tail surfaces of increased span and aspect ratio to which a balanced elevator was now fitted, contributing much more responsive control in the vertical plane. Performance in other parameters was similar to the CL.II, both aircraft being fitted with the same Mercedes D.III engine and the CL.IV retained the communal cockpit provision of the CL.II. Type testing was completed in April 1918, the official report noting its "very favourable climb rate and superlative handling qualities", and the CL.IV was in action from the early summer and served until the conclusion of the war.

Fearsome reputation

By the time the CL.IVs entered the fray German forces were generally on the defensive following the failure of the Spring Offensives of 1918, and the Schlasta units had honed their tactics to a high level of effectiveness. Operating in flights of four to six aircraft, they would relentlessly target and disorganise assembly points if an attack was imminent and then offer support to counter-attacking German troops as the need arose. The CL.IV excelled in these difficult and dangerous missions, becoming popular with its crews and respected by its opponents. Despite the majority of their use being in the ground attack role, CL.IVs were occasionally called upon to operate as escort fighters. Towards the end of the war on sufficiently moonlit nights, they were employed as interceptors to engage

Halberstadt CL.IV

Weight (gross): 1068kg (2355lb)

Dimensions: Length 6.54m (21ft 5in) Wingspan 10.74m (35ft 3in) Height 2.67m (8ft 9in)

Powerplant: One 120kW (160hp) Mercedes D.III 6-cylinder water-cooled inline piston engine

Speed: 165km/h (103mph)

Endurance: 3 hours 15 minutes

Ceiling: 6400 m (21,000ft)

Crew: 2

Armament: One (sometimes two) 7.92mm (0.312in) LMG 08/15 'Spandau' fixed firing forward and one 7.92mm (0.312in) Parabellum MG14 machine gun flexibly mounted in rear cockpit; 10 stick grenades; Up to 5 10kg (22lb) Wurfgranaten 15 trench mortar fragmentation bombs

Allied bombing aircraft returning from missions. Some 450 of these profoundly useful aircraft were ordered from Halberstadt and a further 250 from LFG Roland, deliveries of which were still ongoing as hostilities ended.

Hannover CL-types

Despite never having produced an aircraft before, Hannover developed a formidable two-seater that excelled in the air-to-air and close-support roles.

Hannoversche Wagonfabrik AG was a manufacturer of rolling stock for German railways, but from 1915 it produced aircraft under subcontract to other firms. In 1917, pre-war aviation pioneer Hermann Dorner designed Hannover's first aircraft, the lightweight C.II, retrospectively designated CL.II in the summer of 1917 (there was no C.I, Hannover built the Aviatik C.I under licence and the designation for their first in-house design simply followed this numerical progression).

The CL.II featured a plywood-skinned monocoque fuselage, much like the contemporary Albatros, and was powered by the Argus As.III. An unusual feature was the biplane tail, which was selected to maximise the observer's field of fire directly aft. The observer's position was also raised above the upper fuselage line, affording him an excellent field of fire in the upper hemisphere. Unusually small for a two-seater, enemy fighter pilots often mistook the CL.II for a single seat fighter "until they get to close range, when up pops the Hun gunner from inside his office", as ace James McCudden put it.

In total, 443 examples of the CL.II were built, many by LFG Roland which were referred to as the CL.IIa.

The CL.II passed type tests during July 1917 and was in service with the Schutzstaffeln from October, proving effective and versatile. As well as its primary duties as an escort fighter

Hannover CL.II
Weight (gross): 1081kg (2383lb)
Dimensions: Length: 7.59m (24ft 7in), Wingspan: **11.95m (39ft 2in), Height: 2.8m (9ft 2in)**
Powerplant: One 134kW (180hp) Argus As.III 6-cylinder water-cooled in-line piston engine
Speed: 165km/h (103mph)
Endurance: 3 hours 30 minutes
Ceiling: 7500m
Crew: 2
Armament: One 7.92mm LMG 08/15 'Spandau' fixed firing forward and one 7.92mm (0.312in) Parabellum MG14 machine gun flexibly mounted in rear cockpit

Hannover CL.II

CL.II White 4 of Schlasta 25 wears the distinctive painted lozenge camouflage sported by many Hannovers on the fuselage, the wings, horizontal tail surfaces and wheels being covered with the more conventional printed lozenge fabric. White 4 had a flare rack fitted just below the observer's cockpit.

Hannover CL.III

One of the comparatively rare CL.IIIs, Yellow 1 was on the strength of Feld Abteilung (Artillery) 286b in May 1918. Flown by Heinrich Fichtbauer and Joseph Herz, the aircraft features the early form of the German *balkankreuz* with a non-standard extended horizontal bar on the fuselage cross, as well as a strange but apparently cheerful creature emanating from a beer stein.

for dedicated reconnaissance types, the CL.II was itself employed as a low-altitude tactical reconnaissance platform. Its manoeuvrability was such that its crews could engage enemy single-seaters with confidence, although like the contemporary Halberstadt CL.II, the Hannover was increasingly used for close support on the battlefield rather than as an air superiority asset.

Engine upgrade

In February 1918 a Mercedes D.III powered version with a lighter and narrower fuselage and shorter wingspan designated CL.III passed its type test. The Mercedes engine offered better altitude performance and an order was placed for 200 aircraft. However, the D.III engine was in short supply due to its use in single seat fighters and only 80 C.IIIs were ever built. The balance of the order was completed as the CL.IIIa, which reverted to the Argus As.III engine, and 611 examples were ultimately constructed. The newer aircraft featured redesigned ailerons with aerodynamic balances that improved handling and manoeuvrability, especially at the low levels that the Hannover was increasingly spending most of its time at. Operating with the Schlachtstaffeln in concert with Halberstadt's CL.II and CL.IV, the Hannover CL types enjoyed a similarly successful career, operating effectively against ground targets until the conclusion of hostilities.

The final wartime developments of the CL-type was the CL.IV, a high-altitude escort fighter with a 224Kw (300hp) Maybach engine – of which only five examples were built – and the CL.V that utilised the superb 138kW (185hp) BMW IIIa and possessed a significantly improved performance. In total, 46 CL.Vs had been built by the end of the war but none had entered service.

Surprisingly, Hannover then built another 62 after the armistice until the Treaty of Versailles put an end to all German military aircraft development.

Hannover CL.IIIa

Weight (gross): 1068kg (2355lb)

Dimensions: Length 7.6m (24ft 10in) Wingspan 11.7m (38ft 5in) Height 2.8m (9ft 2in)

Powerplant: One 134kW (180hp) Argus As.III 6-cylinder water-cooled inline piston engine Speed: 165km/h (103mph)

Endurance: 3 hours

Ceiling: 7500m (24,600ft)

Crew: 2

Armament: One (sometimes two) 7.92mm (0.312in) LMG 08/15 'Spandau' fixed firing forward and one 7.92mm (0.312in) Parabellum MG14 machine gun flexibly mounted in rear cockpit

Despite this, a further 14 C.Vs were built for the Norwegian Army in 1923.

Meanwhile the CL.III was developed into a civil aircraft known as the Hawa F.3, and a direct copy of the CL.II became the first aircraft to be built in independent Poland when the CWL SK-1 Słowik flew for the first time in 1919.

With hump-backed profile, odd biplane tail, and slab-sided fuselage, the Hannover two-seaters like this
CL.IIIa could scarcely be considered aesthetically pleasing machines, yet what they lacked in looks they
more than made up for in efficiency. The Hannover CL series were amongst the best two-seat combat
aircraft produced by any nation during the conflict.

Hannover CL.IIIa

The final iteration of the Hannover CL to see service, the CL.IIIa was built in the largest numbers of any
Hannover aircraft. This example, named 'Lena', was on the strength of Bavarian reconnaissance unit
Flieger Abteilung 46b in the latter half of 1918, and wears this unit's distinctive zigzag fuselage band.

Junkers CL.I

The revolutionary Junkers CL.I would have dominated the Schlachtstaffel units had the war continued into 1919, but production difficulties delayed its service entry and it saw only limited action.

Consisting essentially of a stretched two-seat version of the Junkers D.I, the CL.I first flew on 10 December 1917. IdFlieg were profoundly suspicious of all-metal aircraft but reports from Junkers J.I crews testified to the resilience of the metal aircraft to both enemy action and the effects of weather, a serious concern for wooden aircraft. Testing revealed the qualities of the CL.I, the all-metal airframe of which rendered it mostly impervious to small arms fire as well as possessing unparalleled strength for an aircraft of its era. Eventually 63 aircraft were ordered with more imminent, the intention being to replace the Halberstadt CL-types with the new Junkers. Unfortunately IdFlieg dithered over placing an order with Junkers, doubting their capacity to manufacture such large numbers. After considering ordering the aircraft from Linke-Hoffman, IdFlieg finally approved production by

Junkers in collaboration with Fokker. Valuable time was lost and only 43 had been delivered by the armistice, of which a handful reached the Front although it is unlikely any saw action.

Baltic deployment

However, the CL.I did see combat – Freiwilligen Flieger Abteilung 417 of the Kampfgeschwader Sachsenberg flew in support of Freikorps units fighting against Bolshevik insurgents in the Baltic region during 1919. Based in the open in austere cold and damp conditions that would have curtailed the use of any other aircraft, the Junkers machines maintained operations without failure for six months. Three examples of a naval variant were also delivered to the *Kriegsmarine* – CLS.I, which had flown for the first time less than three weeks before the armistice. Consisting of a CL.I with floats

replacing the wheeled undercarriage, the aircraft also featured a large triangular tail fin ahead of the rudder to compensate for the increased side area imparted by the floats. The tail fin would become a familiar design cue on many postwar Junkers' designs, not least the immensely successful F13.

Junkers CL.I

Weight (gross): 1050kg (2310lb)

Dimensions: Length 7.90m (25ft 11in) Wingspan: 12.m (39ft 6in) Height 2.65m (8ft 8in)

Powerplant: One 134kW (180hp) Mercedes D.IIIa 6-cylinder water-cooled inline piston engine

Speed: 161km/h (100mph)

Endurance: 2 hours

Ceiling: 6000m (19,700ft)

Crew: 2

Armament: Two 7.92mm (0.312in) LMG 08/15 'Spandau' fixed firing forward and one 7.92mm (0.312in) Parabellum MG14 machine gun flexibly mounted in rear cockpit

---

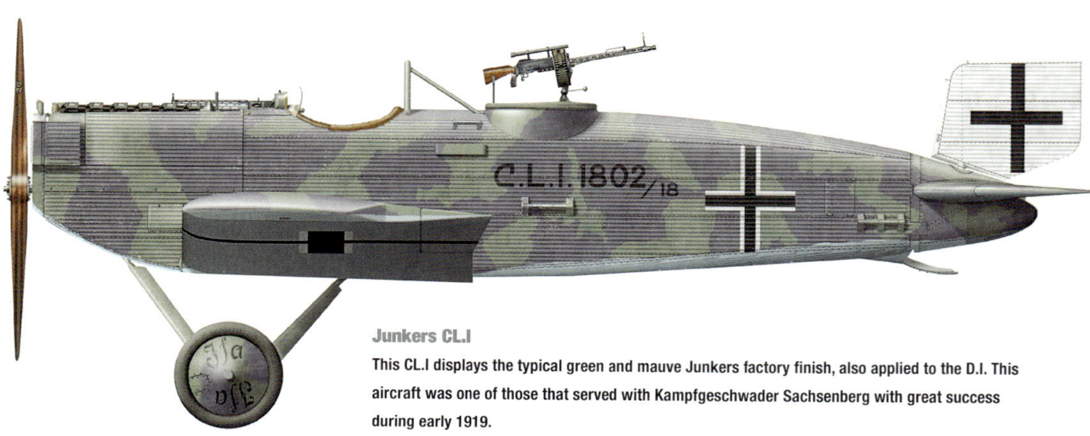

Junkers CL.I

This CL.I displays the typical green and mauve Junkers factory finish, also applied to the D.I. This aircraft was one of those that served with Kampfgeschwader Sachsenberg with great success during early 1919.

Albatros J.I & J.II

Albatros developed an armoured version of its C.XII to produce the close-support J.I. Despite marginal performance, the J.I was an effective machine and served until the end of the war.

Developed specifically for low-altitude ground attacks and battlefield reconnaissance, the J.I utilised the wings and tail of the C.XII and married it to a new fuselage featuring a crew compartment protected by 490kg (1080lb) of nickel-steel armour plate. This resulted in an aircraft some 350kg (772lb) heavier than the C.XII, yet the J.I had to make do with the Benz Bz.IV officially rated at 147kW (197hp) – although the actual power output was slightly higher due to the low altitudes at which the aircraft invariably operated – as opposed to the 194kW (260hp) Mercedes D.IV of the C.XII. As can be imagined, performance suffered as a result and the J.I could only manage 140km/h (87mph) flat out. Rate of climb was also poor but this was not regarded as a serious impediment for an aircraft intended to operate at low altitude. Offensive armament was optimised for ground attack, with two fixed Spandau machine guns firing downwards at a 45 degree angle through the cockpit floor, augmented by the observer's trainable machine gun and various ad-hoc weapons such as hand grenades.

German Spring offensive

The J.I first saw action during April 1918 at the Battle of the Lys, part of Operation Georgette during the German Spring offensive, and proved effective. Operating in flights of three to six at altitudes of 50–500m (164–1640ft), and escorted by Halberstadt and Hannover CL-types, the J.I.s proved deadly against trenches, gun positions and transport columns. Popular with its crews due to its relative invulnerability and very good visibility from the cockpit, the 240 or so built saw intensive action until hostilities ceased. An improved version, the J.II, featured smaller overall dimensions, a more powerful 162kW (217hp) Benz Bz.IVa engine and more armour protection, but was even slower and less manoeuvrable than the J.I. In service from June 1918, around 90 J.IIs were built.

Albatros J.I

Weight (gross): 1808kg (3986lb)
Dimensions: Length 8.8m (28ft 10in) Wingspan 14.14m (46ft 5in) Height 3.3m (11ft)
Powerplant: One 147kW (197hp) Benz Bz.IV 6-cylinder water-cooled inline piston engine
Speed: 140 km/h (87mph)
Endurance: 3 hours 15 minutes
Ceiling: 4500m (14,765ft)
Crew: 2
Armament: Two 7.92mm (0.312in) LMG 08/15 'Spandau' fixed firing forwards and downwards and one 7.92mm (0.312in) Parabellum MG14 machine gun flexibly mounted in rear cockpit

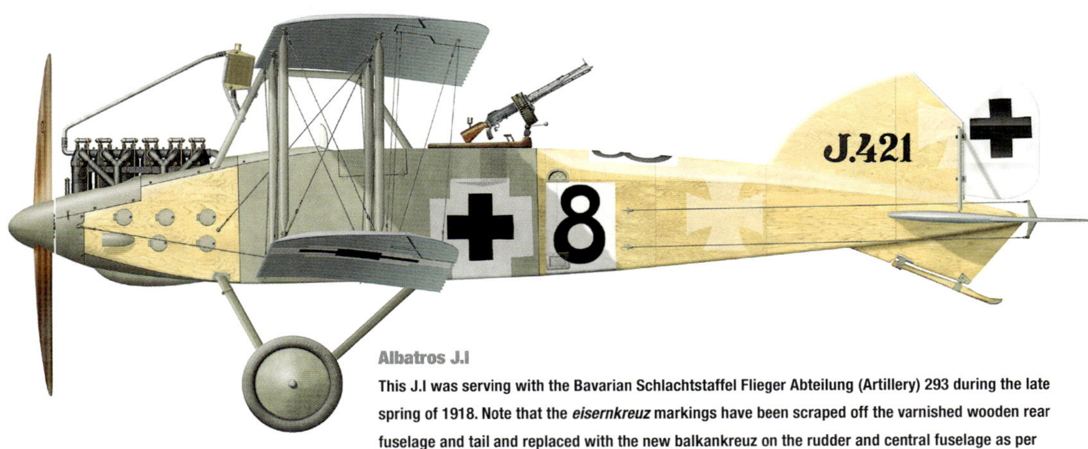

Albatros J.I

This J.I was serving with the Bavarian Schlachtstaffel Flieger Abteilung (Artillery) 293 during the late spring of 1918. Note that the *eisernkreuz* markings have been scraped off the varnished wooden rear fuselage and tail and replaced with the new balkankreuz on the rudder and central fuselage as per IdFlieg's directive of 20 March 1918.

Junkers J.I

As the first mass-produced all-metal aircraft, the Junkers J.I was one of history's most significant aircraft as well as proving highly effective in service. Despite this the 'Furniture Van' remains little known.

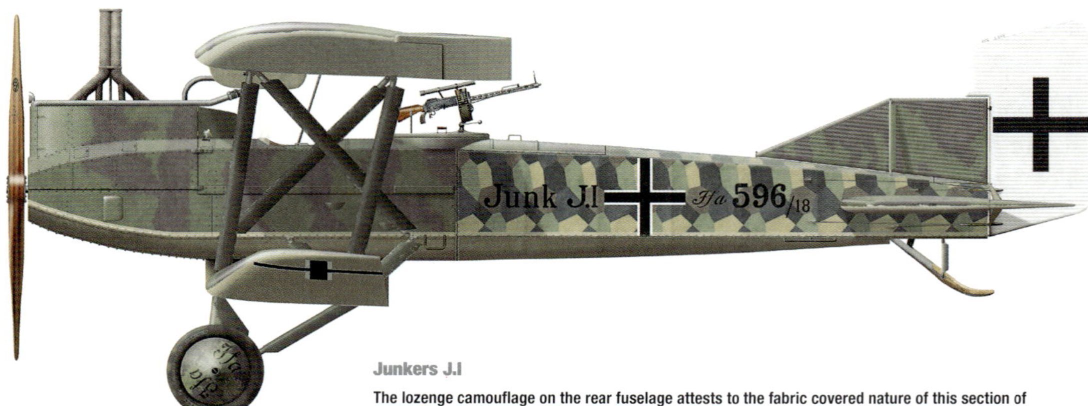

Junkers J.I

The lozenge camouflage on the rear fuselage attests to the fabric covered nature of this section of the airframe. The only other non-metal part of the airframe was the wooden tail skid. This example was from the first batch of aircraft ordered in June 1918 and was discovered by Canadian troops abandoned on its airfield after the armistice.

Engineer, academic and inventor Professor Hugo Junkers founded a company to build boilers and gas stoves in 1892, but his attention shifted to aircraft in 1909 when his company supplied corrugated metal wings for an all-metal canard aircraft designed by Hans Reissner, a fellow professor at the University of Aachen. This was the world's first all-metal aircraft and the experience of working with Reissner seems to have inspired Junkers to enter the realm of aviation with considerable enthusiasm. The development of an all-steel monoplane, the J1, led to IdFlieg contracting Junkers to pursue the all-metal concept and the Junkers J4, intended for ground attack and army cooperation, would be the result, designated J.I by IdFlieg.

Development of the unsuccessful J2 fighter, which was fast but overweight, had led Junkers to abandon electrical steel in favour of duralumin as the main material for the construction of further aircraft designs. Considerable attention had been given to the survivability of the J4 and the heart of the design was a 470kg (1036lb) nickel-steel armoured capsule, 5mm (0.2in) thick, enclosing the engine, fuel tanks and crew compartment, terminating in a solid bulkhead to protect from attack from astern. Push rods were used for operating the control surfaces in place of the more usual cables, as these were less susceptible to damage from ground fire, and the main fuel tank was divided in two to provide a measure of redundancy in the event of it being holed. The rest of the aircraft consisted of an alloy tube frame to which a skin of corrugated duralumin sheet was attached, the rear fuselage and fin and rudder being fabric covered – though later production machines utilised corrugated duralumin skin to cover these areas too. Somewhat ironically, given the aircraft's imperviousness to gunfire, great care had to be exercised when handling the aircraft on the ground as the 0.19mm (0.007in) thick alloy skin of the wings could be very easily dented.

Crew protection

The aircraft flew for the first time on 28 January 1917, yet production was slow and the first examples only reached the Front in August of the same year. Nicknamed the 'Furniture Van' on account of its ponderous handling, once crews adapted themselves to the peculiarities of this revolutionary aircraft it became very popular, not least due to the almost unbelievable level of protection it afforded its occupants. The J.I was not totally invulnerable, however, though many sources claim that none were lost to enemy action, one example was shot down by Sopwith Camel ace Donald Maclaren in March 1918 and

another was brought down by a French anti-aircraft battery firing armour piercing rounds. Nonetheless, these losses were very much the exception to the rule and a J.I crew member could reasonably consider himself very fortunate indeed.

The J.Is were heavily committed to the German Spring offensives of 1918 and operated in the highly dangerous role of 'contact patrol' that consisted primarily of flying at low altitude to establish the course of the front line and any alterations, the observer reporting the position to the rear command post by radio. Ammunition and rations were dropped to machine-gun nests, trench systems and outposts that had been cut off and could not otherwise be supplied and enemy positions were attacked, initially with downward-pointing machine guns – but these proved practically impossible to aim and were repositioned as conventional forward-firing weapons.

In total 227 were built and the aircraft continued in service until the end of hostilities.

Junkers J.I

Weight (gross): 2140kg (4718lb)
Dimensions: Length 9.1m (29ft 10in) Wingspan 16m (52ft 6in) Height 3.4m (11ft 2in)
Powerplant: One 150kW (200hp) Benz Bz.IV 6-cylinder water-cooled inline piston engine
Speed: 155km/h (97mph)
Endurance: 2 hours
Ceiling: 4000m (13,123ft)
Crew: 2
Armament: Two 7.92mm (0.312in) LMG 08/15 'Spandau' fixed firing forward and one 7.92mm (0.312in) Parabellum MG14 machine gun flexibly mounted in rear cockpit

Junkers J.I

The J.I was produced by Junkers in collaboration with Fokker and the 'Jfa' abbreviation for 'Junkers-Fokker Werke AG' appears prominently before the serial number and on the wheel discs. This J.I was damaged in a crash landing during September 1918 but was later repaired and returned to service.

G 107

BOMBERS & SEAPLANES

The use of strategic bombers on civilian targets led to the Germans being branded 'baby killers' by their opponents, but the campaigns waged by both Zeppelins and conventional aircraft pointed the way towards the horrors of total warfare to come in the next conflict. This chapter also includes the relatively few maritime types operated in numbers by the Imperial Navy.

This chapter includes the following aircraft:
- AEG G.I to G.IV
- Zeppelin airships
- Friedrichshafen G.III & G.IV
- Gotha G.I
- Gotha G.II to G.V
- Rumpler G series
- Siemens-Schuckert R series
- Zeppelin-Staaken R series
- Gotha WD series
- Albatros W series
- Hansa-Brandenburg W.12
- Hansa-Brandenburg W.19
- Hansa-Brandenburg W.29

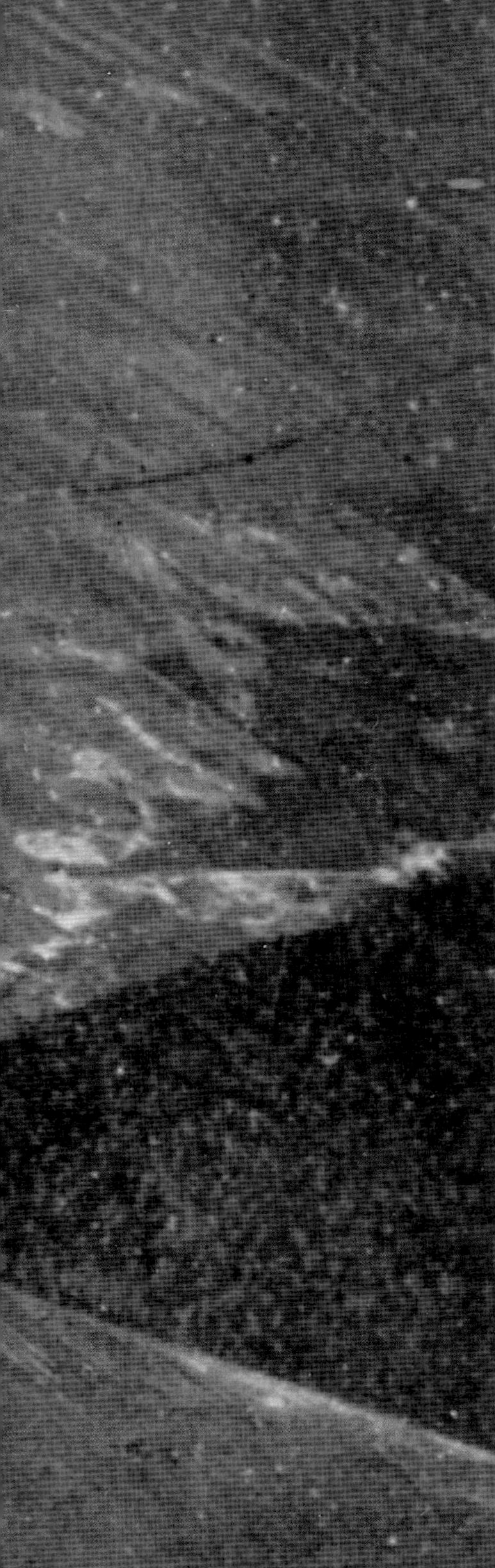

The Rumpler G.II was one of the first large aircraft operated by the Germans and was used for reconnaissance, bombing and even as an escort for smaller machines. This example was operated by Kagohl II when it was photographed over the Eastern Front in 1916.

AEG G.I to G.IV

After building a series of generally inadequate large aircraft, AEG delivered the G.IV that although deficient in range was both rugged and easy to fly.

The first of AEG's large kampfflugzeug aircraft, the K.I, first flew in early 1915. This was intended to operate as a form of aerial cruiser to engage and destroy enemy aircraft but also possessing the ability to carry a bombload.

AEG G.I

The K.I was sufficiently promising for a developed version, the G.I, to be sent to the Front for assessment under operational conditions. IdFlieg then ordered a small production run of the improved G.II, 27 of which were built. Combat experience revealed the Kampfflugzeug concept to be a failure, as the G.II was too slow and cumbersome to catch enemy aircraft. It was officially noted by IdFlieg that it "had fared poorly in air combat, but

had been successfully employed as a bomber in squadron strength".

After a few attempts at escort duty the G.II and the slightly improved G.III, of which 45 were built, were both used exclusively in the bombing role.

All-metal construction

The G.IV was a more formidable development, utilising the six-cylinder Mercedes D.IVa that replaced the unreliable eight-cylinder Mercedes D.IV of the G.III. Featuring a welded tube all-metal fuselage frame, the G.IV was much stronger than contemporary bombers such as the Gotha and was considerably easier to fly. Its range was the shortest of its contemporaries so the G.IV was mainly employed in the tactical role, both by day and by night, although as

AEG G.IV

Weight: 3630kg (8003lb)

Dimensions: Length 9.7m (31ft 10in) Wingspan 18.4m (60ft 4in) Height 3.9m (12ft 10in)

Powerplant: Two 190kW (260hp) Mercedes D.IVa 6-cylinder water-cooled inline piston engines

Speed: 165km/h (103mph)

Endurance: 4–5 hours

Ceiling: 4500m (14,800ft)

Crew: 3

Armament: One 7.92mm (0.312in) Parabellum MG14 machine gun flexibly mounted in nose position, one 7.92mm (0.312in) Parabellum MG14 machine gun flexibly mounted in rear cockpit, which could also be fired through a ventral hatch; up to 400kg (880lb) bombload

AEG G.IVb

'White 21' was a G.IVb aircraft that were produced as part of a production order for standard G.IVs in 1916. The G.IVb had an upper wing of greater span to allow it to carry a single 1000kg bomb but very few were produced.

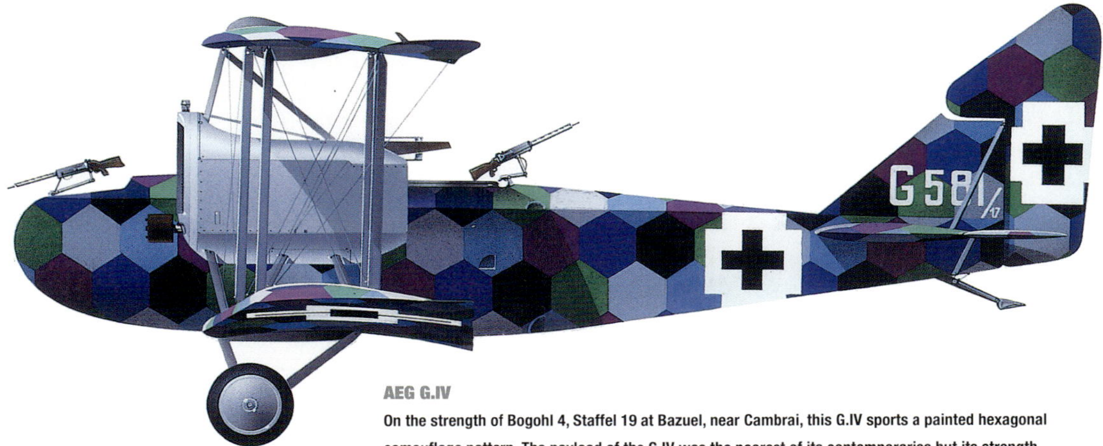

AEG G.IV

On the strength of Bogohl 4, Staffel 19 at Bazuel, near Cambrai, this G.IV sports a painted hexagonal camouflage pattern. The payload of the G.IV was the poorest of its contemporaries but its strength and ease of handling endeared it to its crews.

the conflict progressed the G.IV found itself employed near exclusively on nocturnal raids.

Well-armed defensively, the G.IV also featured radio equipment and heated suits for the crew. In total 320 were built before production switched to the G.V but this never became operational.

After the war several G.Vs were converted as six-passenger airliners for Deutsche Luft-Reederei, proving unusually successful for a Great War military aircraft conversion.

AEG G.IV

Seen here 'bombed up', 'White VII' is likely the most spectacularly marked AEG G.IV to see service, appearing in this scheme in 1918 with an unknown unit. In addition to the ferocious shark's mouth, the aircraft features printed lozenge camouflage overpainted with black for the nocturnal role.

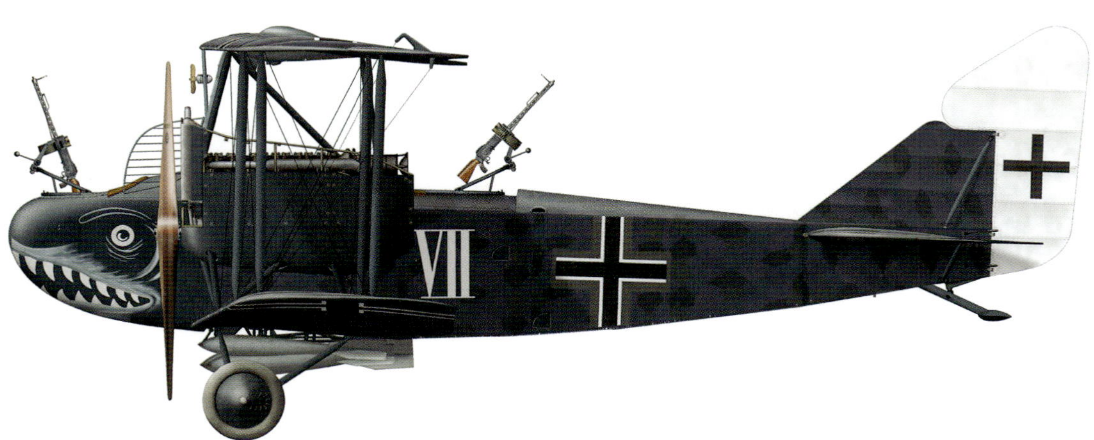

Friedrichshafen G.III & G.IV

Not as famous as its Gotha stablemate, largely because it never bombed London, the Friedrichshafen G-types were arguably superior and built in greater numbers.

Friedrichshafen built one G.I to the same Kampfflugzeug specification as AEG's G.I. The Friedrichshafen aircraft was found to be similarly worthy of further development and production of an initial six G.IIs with two Benz Bz.IV engines was initiated in late 1915. Eventually 35 of the G.II series were built before production switched to the Mercedes D.IVa-powered G.III.

Improved handling

The G.III was built in vast numbers for such a large aircraft, 734 rolling off the production line as well as 138 of the very similar G.IV. Entering service in early 1917, it proved popular. Compared to its contemporaries, the G.III was a much easier aircraft to fly, and particularly land, than the Gotha types and it was very strong. It also

possessed a greater load carrying capability than the Gotha – up to 1000kg (2204lb) of bombs could be carried, though around 600kg (1323lb) was a typical operational load. The G.III was employed as both a tactical and strategic bomber, mainly on the Western Front, and although it never attacked London, Friedrichshafens are known to have bombed Paris.

Vulnerability to attack from below led to the adoption of the 'Gotha tunnel' on the G.III, allowing the rear gunner to fire downwards and to the rear. This variant was known as the G.IIIa and also featured a biplane 'box' tail that improved handling, particularly when flown on one engine. The G.IIIb was a further modification that allowed the crew access to each other's positions during flight. The

final variant, the G.IV, featured tractor propellers and dispensed with the nose gunner. Although built in considerable numbers, it is not definitively known if this version saw operational service before the armistice.

Friedrichshafen G.III

Weight: (gross) 3795kg (8367lb)

Dimensions: Length 12.65m (41ft 6in) Wingspan 23.85m (78ft 3in) Height 4.14m (13ft 7in)

Powerplant: Two 190kW (260hp) Mercedes D.IVa 6-cylinder water-cooled inline piston engines

Speed: 140km/h (87mph)

Endurance: 5 hours

Ceiling: 4500m (14,800ft)

Crew: 3

Armament: One 7.92mm (0.312in) Parabellum MG14 machine gun flexibly mounted in nose position, one 7.92mm (0.312in) Parabellum MG14 machine gun flexibly mounted in rear cockpit; up to 1000kg (2200lb) bomb load

---

Despite being built in greater numbers, the Frieridchshafen G.III never achieved the fame of the contemporary but less capable Gotha. After the war several Friedrichshafens were operated by the Government and the commercial airline Deutsche Luft-Reederei (DLR) until the 1919 Treaty of Versailles brought all such operations to an abrupt end.

Gotha G.I

The distinctive Gotha G.I was the first bomber to be produced by Gothaer Waggonfabrik AG, and although not a spectacular success it gave the company valuable experience in producing large aircraft.

The G.I was designed by Oskar Ursinus, founder and editor of Flugsport magazine, and featured an unusual configuration in which the fuselage was attached to the top wing with the engine nacelles mounted below it. These were very close together, propellers almost touching, the purpose of this arrangement being to minimise asymmetric handling issues in the event of an engine failure. The two propellers were also arranged to turn in opposite directions, cancelling torque issues. The position of the fuselage also conferred an exceptional field of fire for the gunners, the downside being that this arrangement would prove invariably fatal to the crew in the event of a crash landing.

Fighter vulnerability

The prototype of Ursinus' remarkable aircraft was built by the unit he was serving in, Fliegerersatz Abteilung 3 ('Aviator Replacement Unit 3') and received the IdFlieg designation B.1092/14 although it was generally referred to as the FU for Friedel-Ursinus (Major Helmut Friedel was Oskar Ursinus' superior officer). The FU was tested with a reconnaissance unit on the Russian Front and despite being found to be difficult to fly, underpowered and structurally weak, on 1 April 1915 IdFlieg ordered Gotha to produce 18 aircraft. The production aircraft were used mostly in the East and proved adequate although vulnerable to fighter attack even at this early stage of aerial combat. In service the majority of missions were reconnaissance and patrols, bombing sorties taking place only very occasionally.

A single seaplane version, the UWD, was also built. Nicknamed the Trojanisches Pferd ('Trojan Horse'), the single UWD led a busy life, including a successful bombing raid on military installations at Dover during 1916. It subsequently spent most of its life as a trainer for torpedo aircraft crews.

Gotha G.I

Weight: (gross) 2966kg (6539lb)
Dimensions: Length 12.1m (39ft 8in) Wingspan 20.3m (66ft 7in) Height 3.9m (12ft 10in)
Powerplant: Two 110kW (150hp) Benz Bz.III 6-cylinder water-cooled inline piston engines
Speed: 130km/h (81mph)
Endurance: 4 hours
Ceiling: 2750m (9020ft)
Crew: 3
Armament: One 7.92mm (0.312in) Parabellum MG14 machine gun flexibly mounted in nose position, one 7.92mm (0.312in) Parabellum MG14 machine gun flexibly mounted in centre cockpit; up to 350kg (770lb) bombload

---

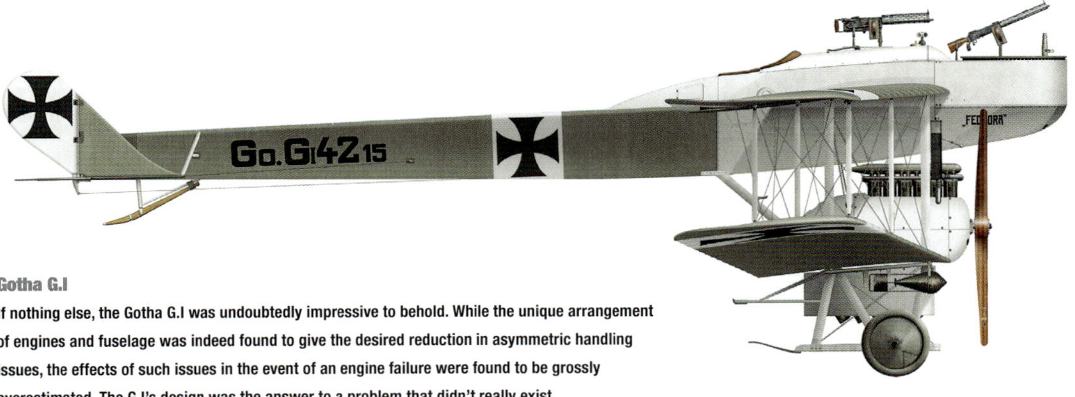

Gotha G.I

If nothing else, the Gotha G.I was undoubtedly impressive to behold. While the unique arrangement of engines and fuselage was indeed found to give the desired reduction in asymmetric handling issues, the effects of such issues in the event of an engine failure were found to be grossly overestimated. The G.I's design was the answer to a problem that didn't really exist.

Gotha G.II to G.V

The Gotha bombers were the world's first mass-produced large aircraft. Their strategic bombing campaign against London had a psychological effect out of all proportion to the damage inflicted.

Hans Burkhard, Gotha's chief designer, decided to improve Okar Ursinus' G.I by altering it into a more conventional configuration. Using a crashed G.I airframe, Burkhard relocated the fuselage to a more conventional location on the lower wing, moving the engine nacelles further outboard to accommodate this change. The engines were Mercedes D.IV eight-cylinder units with pusher propellers rather than the tractor arrangement of the G.I. The new Gotha G.II flew in March 1916 and initially could not carry the bombload specified by IdFlieg, though this was solved relatively simply by increasing the wingspan. The aircraft also utilised an unusual quadricycle landing gear with wheels at the Front and rear of each engine nacelle. However, this was found to result in an long and

near-uncontrollable landing run and was changed to a conventional tailskid arrangement. In this form the G.II entered limited production with a total of 11 being built.

Balkan deployment

Eight of the G.IIs saw service in the Balkans from August 1916 but details of their combat use is obscure, although it is known that all had been withdrawn by April 1917. The Mercedes D.IV was notoriously prone to crankshaft failure and a change to the six-cylinder Mercedes D.IVa resulted in the Gotha G.III, of which 25 were built, this also featuring a ventral hatch through which the rear

gunner could fire downwards and to the rear. The G.III was also used on the Balkan Front and one managed to hit the Romanian Cernavod Bridge, then Europe's longest, in September 1916 (the bridge survived).

Gun tunnel

The first Gotha to be mass produced was the G.IV. This was basically the same as the G.III but with ailerons on all four wingtips and the Gotha 'gun tunnel', a trough-shaped cut-out in the rear fuselage allowing the rear gunner to fire downwards. This feature was incorporated after gunners experienced difficulty using the ventral hatch on the G.III. The structural integrity lost as a

Gotha G.IV

Weight: 3648kg (8042lb)
Dimensions: Length 12.2m (40ft) Wingspan 23.7m (77ft 9in) Height 3.9m (12ft 10in)
Powerplant: Two 190kW (260hp) Mercedes D.IVa 6-cylinder water-cooled inline piston engines
Speed: 135km/h (84mph)
Endurance: 6 hours
Ceiling: 5000m (16,400ft)
Crew: 3
Armament: One 7.92mm (0.312in) Parabellum MG14 machine gun flexibly mounted in nose position, one 7.92mm (0.312in) Parabellum MG14 machine gun flexibly mounted in dorsal/ventral position; up to 500kg (1100lb) bombload

Gotha G.IV

Detail differences abounded between sub-contractors and this bombed-up LVG-built G.IV features extra tail struts. The name 'Morotas' refers to its crew – Lt. Mons, Lt Roland and Herman Tasche. This aircraft crashed into a Belgian farmhouse on the night of 10–11 November 1917. Herman Tasche later survived being shot down in a G.V over England.

result of the gun tunnel was regained by skinning the rear fuselage in plywood.

London raids

Entering service in the autumn of 1916, the Gotha allowed strategic raids on London to be mounted by fixed-wing aircraft for the first time. The initial raid was flown on 13 June 1917 and was the first time London had been bombed by day. The raid proved spectacularly effective – all 20 aircraft dispatched returned safely and the 14 that bombed London caused an estimated £125,953 of damage and killed 162 people.

The second was on 7 July, resulting in 54 more civilian deaths and so incensed the population that there were riots in East London with mobs attacking immigrant shops and houses. Four days later, George V officially changed the royal family's name to Windsor so as to remove the obvious German bombing connotations of Saxe-Coburg-Gotha.

The London raids were flown by night from September due to improving defences and continued until May 1918, with 24 Gothas being shot down in the process before

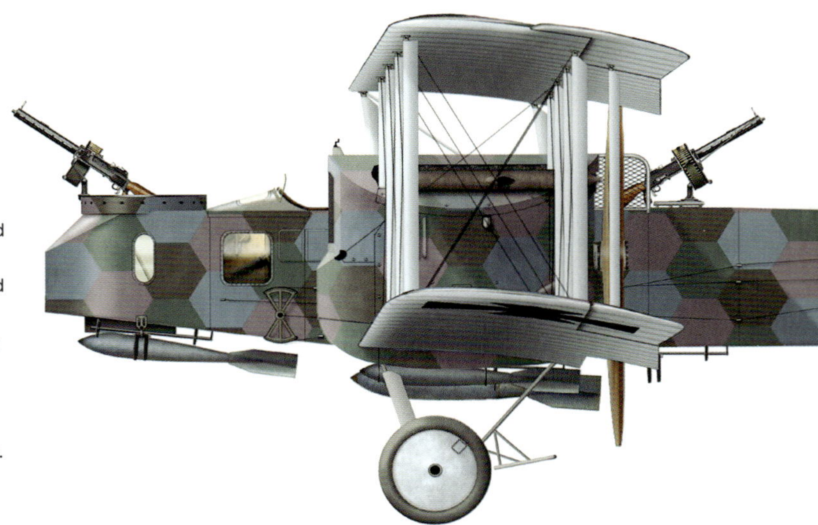

Gotha G.IV

Weight: (gross) 3648kg (8042lb)

Dimensions: Length 12.2m (40ft) Wingspan 23.7m (77ft 9in) Height 3.9m (12ft 10in)

Powerplant: Two 190kW (260hp) Mercedes D.IVa 6-cylinder water-cooled inline piston engines

Speed: 135km/h (84mph)

Endurance: 6 hours

Ceiling: 5000m (16,400ft)

Crew: 3

Armament: One 7.92mm (0.312in) Parabellum MG14 machine gun flexibly mounted in nose position, one 7.92mm (0.312in) Parabellum MG14 machine gun flexibly mounted in dorsal/ventral position; up to 500kg (1100lb) bombload

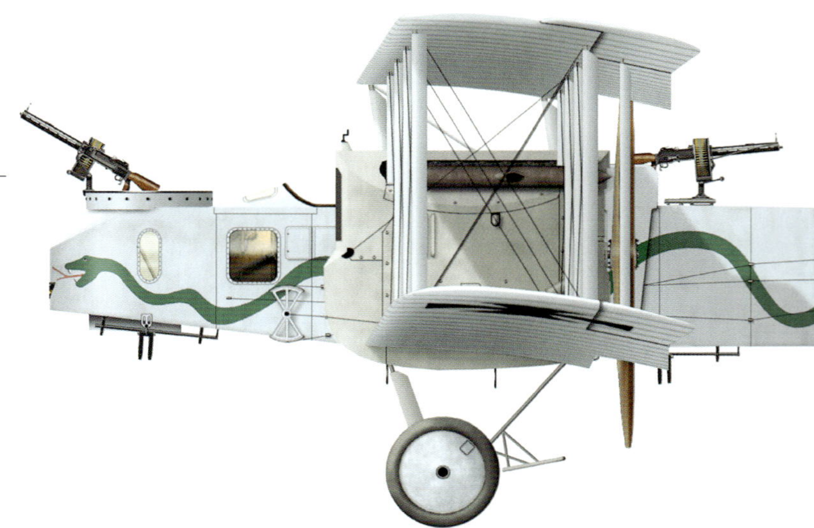

switching to other targets. From August 1917, the G.IV had been joined in the London raids by the generally similar G.V, which differed primarily in having the fuel tanks relocated from the engine nacelles to a less vulnerable position in the fuselage.

Both the G.IV and G.V became tail-heavy and difficult to control once the bombs had been released and 36 were lost in landing accidents, a third more than to enemy action.

A total of 230 G.IVs and 205 G.Vs were built, all but one G.IV (which was captured by the Poles) being destroyed in 1919 in accordance with the Treaty of Versailles.

Gotha G.IV

Siemens-Schuckert Werke built 80 Gotha G.IVs under licence during 1916 including this example, as can be seen by the SSW suffix to the serial number. Siemens-Schuckert built G.IVs, as well as some built by LVG, were camouflaged at the factory in this distinctive hexagonal pattern of greys, dark blues and mauve tones.

Gotha G.IV

This G.IV, in the blue-grey finish typical of Kagohl 3 (the 'Englandgeschwader'), is believed to have been the 'Serpent Machine' of Walter Aschoff, lavishly decorated with snakes. Aschoff was shot down over Britain but survived to write a memoir of his experiences called *Londonflüge 1917.*

Rumpler G series

Rumpler produced a selection of large bombing aircraft from the beginning of 1915 to early 1917. The Rumpler G-types proved moderately successful but were never built in quantity.

The Rumpler G.III was never a common aircraft but it did enjoy a longevity of service unusual for an aircraft of its vintage. The twin-engine Rumplers proved dependable and popular with their crews, operating primarily in the reconnaissance role.

Several manufacturers prepared designs for the large kampfflugzeug requirement, Rumpler submitting the 4A 15 that appeared in March 1915. The 4A 15 caused something of a stir when it carried 10 people to a height of 3200m (10,500ft) on 15 March 1915, followed later the same day by reaching 1800m (5905ft) while carrying 16 people, an enormous amount by contemporary standards. Although destroyed as the result of a carburettor fire in April, Rumpler was already working on an improved version, the 5A, 15 of which entered very limited production in September 1915, although only four were built as the G.I.

Reconnaissance role

In November 1915 IdFlieg ordered 24 of the improved G.II, which differed mainly in the substitution of Benz Bv.IV engines for the Bz.IIIs of the G.I. The aircraft initially possessed a marked tail-heaviness, but this was solved by introducing distinctive cut-outs to the trailing edge of the lower wing. The G.IIs saw most of their service with Kagohl 1 and 2 on both the Eastern and Western Fronts and proved quite successful, being mostly used as a long-range reconnaissance platform though the G.II also employed its relatively heavy armament to operate as an escort for smaller C-type aircraft on bombing raids.

A final variant, the G.III, utilised the Mercedes D.IVa for increased performance and arrived at Kagohl 2 in December 1916. The G.III again suffered from tail heaviness, this time cured by moving the wings backwards. In total 30 G.IIIs were built and some served until at least March 1918. By this time Rumpler had stopped working on large aircraft to concentrate on its highly successful series of two-seater aircraft. A final heavy bomber design,

Rumpler G.III

Weight: (gross) 3620kg (7964lb)
Dimensions: Length 12m (39ft 4in) Wingspan 19.3m (63ft 4in) Height 4.5m (19ft 2in)
Powerplant: Two 190kW (260hp) Mercedes D.IVa 6-cylinder water-cooled inline piston engine
Speed: 165km/h (103mph)
Range: 700km (440 miles)
Ceiling: 5000m (16,000ft)
Crew: 3
Armament: One 7.92mm (0.312in) Parabellum MG14 machine gun flexibly mounted in nose position, one 7.92mm (0.312in) Parabellum MG14 machine gun flexibly mounted in rear cockpit; up to 250kg (550lb) bombload

the 6G4 of February 1918, had proved to have serious control problems and with better aircraft available from other manufacturers, its development was abandoned.

Siemens-Schuckert R series

Siemens-Schuckert flew a series of seven giant aircraft between May 1915 and January 1917. Despite their unusual layout and enormous size, the Siemens-Schuckert R-types proved surprisingly successful and enjoyed long service lives.

Brothers Franz and Bruno Steffen designed a very large aircraft for Siemens-Schuckert in 1914 that garnered enough interest from IdFlieg for a prototype to be ordered. Completed in spring 1915, the aircraft featured three 112kW (150hp) Benz Bz.III engines buried in the fuselage powering two tractor propellers mounted between the wings via drive shafts.

The rear fuselage was split, the top and bottom sections forming two large triangular-shaped structures that terminated in the tail unit. This highly unusual arrangement provided a clear field of fire directly to the rear of the aircraft for a gun position level with the trailing edge of the wing.

Originally designated G.I, for Grossflugzeug, a change was made to R.I Reisenflugzeug after the aircraft was built. A planned series of seven Siemens-Schuckert R.Is were to be built but ultimately all were sufficiently different to each be designated as a distinct (and unique) aircraft type by IdFlieg: R.I to R.VII.

Design vulnerabilities

The R.I was never taken into service but served as a trainer instead, possibly until the armistice. Due to various delays and mishaps, mostly concerning the 180kW Maybach HS engines initially fitted which were prone to overheating, the R.II and R.III never saw operational service either and were likewise used in the training role once fitted with alternative engines. The first to operate at the Front was the R.VI, which was built with three reliable 150kW (200hp) Benz Bz.IVs, and saw service on the Eastern Front between July 1916 and November 1917 when it was declared obsolete and dismantled. The R.IV, R.V and R.VII also saw front-line service, performing reconnaissance flights and bombing missions until their slow speed and size rendered them too vulnerable to continue on operations and all were relegated to training by the end of 1917.

It is not known when the R series was finally retired but the R.IV was still flying as late as August 1918.

Siemens-Schuckert R.V

Weight: (gross) 6766kg (14,885lb)

Dimensions: Length 17.7m (58ft 1in) Wingspan 34.33m (112ft 8in) Height 4.6m (15ft 8in)

Powerplant: Three 150kW (200hp) Benz Bz.IV 6-cylinder water-cooled inline piston engines

Speed: 132km/h (83mph)

Range: 480km (300 miles)

Ceiling: 3000m (9800ft)

Crew: 4

Armament: One 7.92mm (0.312in) Parabellum MG14 machine gun flexibly mounted in nose position, one 7.92mm (0.312in) Parabellum MG14 machine gun flexibly mounted in rear position at junction of tail booms, one 7.92mm (0.312in) Parabellum MG14 machine gun flexibly mounted in ventral hatch; up to 850kg (1875lb) bombload

---

With its bizarre twin triangular rear structure, engines buried in the fuselage and extensively glazed, enclosed cockpit, the Siemens-Schuckert R-types were amongst the oddest looking aircraft of the conflict. Despite the combination of unorthodox features in one airframe, the three-engined Siemens-Schuckerts proved surprisingly reliable in service.

Zeppelin-Staaken R series

As well as massive airships, the Zeppelin company also built massive aeroplanes. The impressive R.VI was the largest heavier-than-air aircraft of World War I to enter serial production.

Ironically, the development of the largest aeroplane ever to attack the British Isles was inspired by a British newspaper: in 1913 *The Daily Mail* offered a prize of £10,000 for the first aeroplane to fly across the Atlantic. A contender for the prize was being built at one of industrialist Robert Bosch's factories when war halted work. Count Ferdinand Zeppelin, the airship pioneer, persuaded Bosch to continue development, jointly forming the Versuchsbau GmbH at Gotha-Ost (VGO) for such a purpose. The three-engine aircraft had its maiden flight on 11 April 1915, by which time IdFlieg had come up with the Reisenflugzeug designation and accepted the three VGO built giants into service. All three saw action on the Eastern Front, VGO.I initially in Naval hands and VGO.II and III with the Luftskreitkräfte.

Production of the Zeppelin giants moved to Staaken near Berlin during

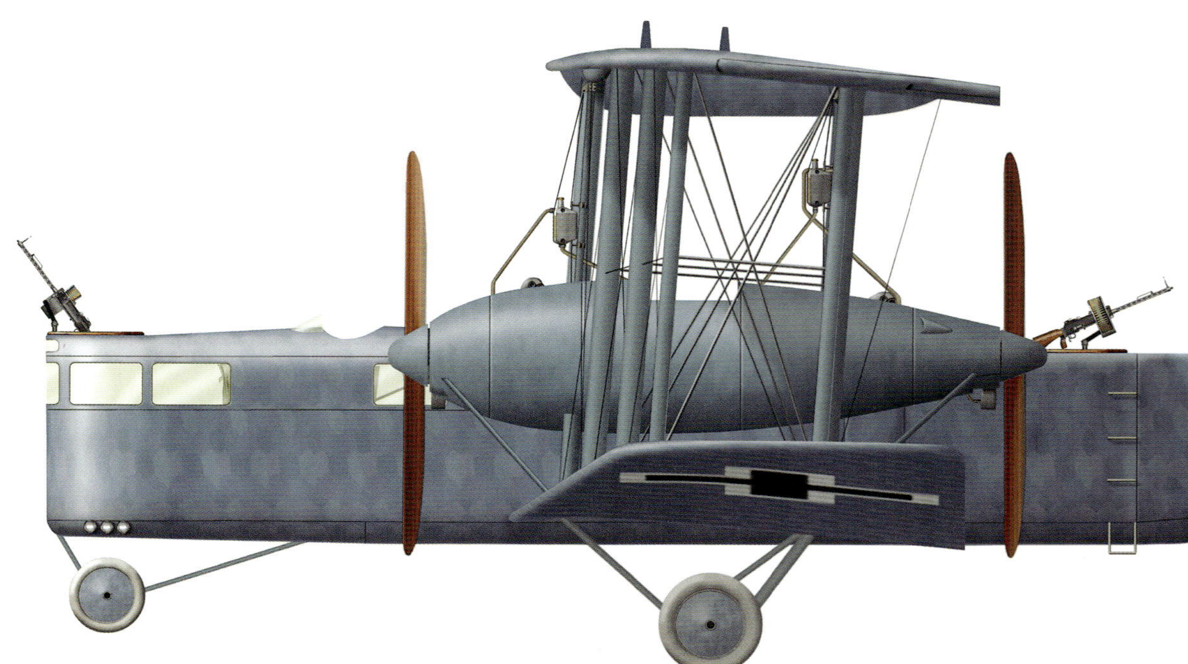

1916, safely distant from any attack by the Allies, and in early 1917 the next giant was rolled out: the Zeppelin-Staaken R.IV. A development of the VGO aircraft, the R.IV featured six engines, four Benz Bz.IVs in pairs (each pair driving a propeller at the rear of the engine nacelle) and two Mercedes D.IIIs mounted side by side in the nose, geared to a single large propeller. The R.IV was destined to be a particularly successful giant, serving initially on the Eastern Front before being transferred to the West.

During a Gotha raid on London on 18 January, the R.IV approached alone from the north east, knocking out the engine of an attacking Bristol F.2b night fighter over Essex before bombing London. One of its bombs hit a print works being used as a bomb shelter, killing 38 and wounding 85, the most casualties caused by a single bomb during the London raids.

London raids

The most important Staaken model was the R.VI of which 18 were constructed, all but six under licence by Albatros, Aviatik and Schütte-Lanz. Dispensing with the nose engine layout and featuring a fully enclosed cabin, the R.VI proved quite successful. After operating on the Eastern Front from September 1917, the R.VIs of Riesenflugzeug-Abteilung (Rfa) 501 moved to Ghent and flew 11 raids on London, dropping 27,190kg (59,944lb) of bombs on British targets until May 1918. They were responsible

Zeppelin-Staaken R.VI

Weight: (gross) 11,848kg (26,120lb)

Dimensions: Length 22.1m (72ft 6in) Wingspan 42.2m (138ft 5in) Height 6.3m (20ft 8in)

Powerplant: Four 183kW (245hp) Maybach Mb.IVa high compression 6-cylinder water-cooled inline piston engines

Speed: 135km/h (84mph)

Endurance: 7–10 hours

Ceiling: 4320m (14,170ft)

Crew: 10

Armament: One 7.92mm (0.312in) Parabellum MG14 machine gun flexibly mounted in nose, dorsal and ventral positions; up to 2000kg (4409lb) bombload

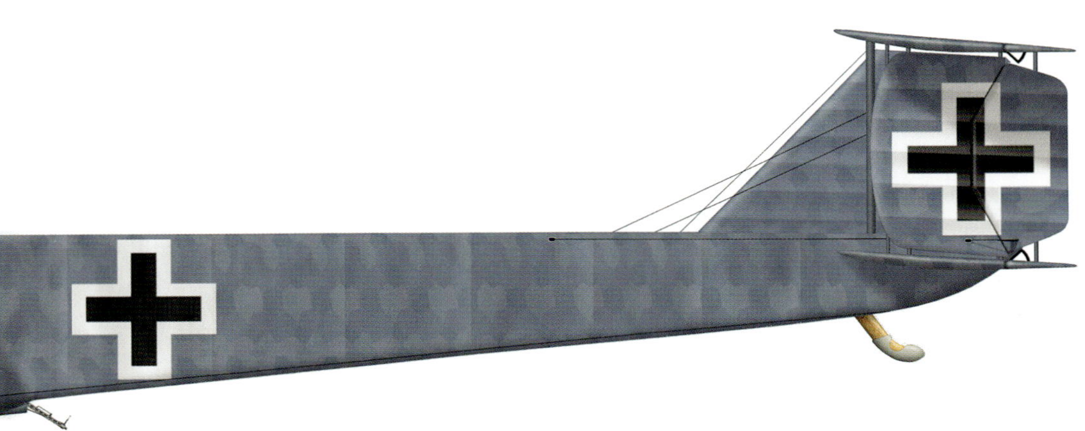

Zeppelin-Staaken R.VI

The giant Zeppelin-Staakens were the largest conventional aircraft to see service during World War I. This example of the R.VI, the most numerous model, was built under licence by Aviatik and served with Rfa 500 on the Western Front during the summer of 1918.

Zeppelin-Staaken R.IV

The R.IV was the only one of the early Zeppelin giants to survive the war. Featuring a gun position in each nacelle as well as one on the training edge of the upper wing, the R.IV was an unusually well-defended aircraft.

for delivering the first 1000kg (2205lb) bombs dropped on the British Isles and none was lost to enemy action, although two crashed on their return to base. A further seven Zeppelin-Staakens were built during the war, but no more than three of any specific design, and only the five engine R.XIVs appear to have seen operational service, one of them being shot down by a Sopwith Camel on the night of 10 August 1918.

Post-war deployments

In the fraught political situation immediately following the war, four of the remaining giants were employed to fly currency from the German government to support independent Ukraine against invading Russian and Polish armies. One was shot down by Polish border troops, one was interned in Vienna and another was confiscated by Romania after it force landed in a cornfield after being blown off course. Much was made of the potential of the Reisenflugzeug as civil aircraft but apart from a few publicity flights, one of the R.XVIs even appearing in a feature film, the aircraft proved too expensive and complicated to keep

The people clambering about on this brand new Zeppelin-Staaken R.VI in 1917 give an inkling as to the incredible size of the aircraft. The man on the wing is standing next to the side door which allowed the crew to climb across from the fuselage to the engine nacelles in flight, or vice versa, not a journey for the faint hearted.

flying. Three Zeppelin floatplanes were also built, based on the R.VI, and despite never seeing military service two of them were used to fly 24 passengers at a time on joyrides from the Havel Lakes near Berlin.

Gotha WD series

Despite gaining fame as a builder of heavy bombers, Gotha also produced a series of twin engine seaplanes, one of which would become the first German aircraft to sink an enemy ship by torpedo attack.

The design that would become Gotha's most successful seaplane started life as the WD.7, a twin-engine, two-float seaplane powered by two Mercedes D.II engines turning pusher propellers. Originally intended as a three-seat kampfflugzeug to attack enemy aircraft, on 5 April 1916 the prototype caught fire on its first combat mission and its crew and their partly burnt aircraft were captured by the French. Undeterred, the Navy ordered a further seven aircraft with lower powered engines that were used as trainers.

Meanwhile, Gotha had developed the WD.11 torpedo bomber, a larger aircraft powered by Mercedes D.III engines, and 17 were built. Despite being considered underpowered the type did become operational in the North and Baltic Seas. Torpedo attacks were difficult, the aircraft had to maintain a height of 5m (16ft) for torpedo release, which was ascertained by lowering a rod from the fuselage until it dragged in the water. This,

combined with the extreme vulnerability of the aircraft during its attack, led to its modification to carry 10 50kg (110lb) bombs or a single anti-shipping mine instead. However, the torpedo aircraft did achieve some success. On 14 June 1917, for example, a WD.11 torpedoed and sank the S.S. *Kankakee* in the Thames estuary.

W.D.14

The improved WD.14, of which 52 were delivered, was also intended to be a torpedo bomber but the difficult combat experience of the WD.11 resulted in most WD.14s being repurposed as a long-range maritime reconnaissance and patrol machine. Initial production machines featured a nose gun position but later this was deleted and pilot and observer sat side by side in the cockpit, greatly aiding communication. With a jettisonable fuel tank replacing the torpedo the WD.14 could stay aloft for an impressive eight-and-a-half hours.

Gotha WD.14

Weight: (gross) 4642kg (10,234lb)
Dimensions: Length 14.45m (47ft 5in) Wingspan 25.5m (83ft 8in) Height 5m (16ft 5in)
Powerplant: Two 150kW (200 hp) Benz Bz.IV 6-cylinder water-cooled inline piston engine
Speed: 130km/h (81mph)
Endurance: 4 hours in torpedo bomber role, 8.5 hours with auxiliary fuel tank
Ceiling: 3200m (10,500ft)
Crew: 3
Armament: One 7.92mm (0.312in) Parabellum MG14 machine gun flexibly mounted in nose position, one 7.92mm Parabellum MG14 machine gun flexibly mounted in rear cockpit; one 725kg (1598lb) torpedo

---

An imposing aircraft, the twin engine WD.11 was neither commonplace nor particularly satisfactory in its intended role. Nevertheless, the WD.11 did achieve a significant milestone with the first successful German air-launched torpedo attack, though the use of this weapon was found to be prohibitively difficult and perilous for crews.

Albatros W series

A seaplane derivative of the D.I, the Albatros W.4 suffered many teething troubles but ultimately proved to be a successful seaplane fighter, remaining in front-line service until the summer of 1918.

When the Navy requested a single-seat floatplane fighter, Albatros responded with a derivative of their D.I fighter that had been proving extremely successful on land. The W.4 was necessarily slower and heavier due to the weight and drag of the floats, and it required wings of greater area to supply extra lift. The horizontal tail surfaces also needed to be enlarged to cope with the larger wings and initial testing revealed the aircraft to be tail heavy. This problem was solved by decreasing the wing stagger, which also happily increased climb rate and maximum speed.

Extensive modifications

The first batch of W.4s was delivered during September 1916 and the aircraft would be ordered in a succession of small batches until December 1917, by which time 118 had been built. Initial service revealed that the aircraft had excellent performance for a floatplane but there were headaches: the wooden floats were frequently modified due to the need to improve both strength and seaworthiness, the lower wings had to be waterproofed after damage was found in the wing spars, the fuselage-mounted 'ear' radiators were prone to overheating in warm weather and were replaced by an aerofoil-shaped radiator in the wing, and inadequate handling saw the addition of ailerons on all four wingtips.

The problems were eventually overcome and the W.4 enjoyed quite a long period of service although the floats did have to be regularly replaced. The most successful pilot of the W.4 was Viktor Schulz, who managed to score three victories with the aircraft, a creditable achievement as the theatres in which it operated were in no way as target rich as the Western Front. Service use wound down during 1918 but there were still four operational in the North Sea during August with another five in Turkey.

Albatros W.4 (late production)

Weight: (gross):1064kg (2346lb)
Dimensions: Length 8.4m (27ft 7in) Wingspan 9.5m (31ft 2in) Height 3.6m (11ft 10in)
Powerplant: One 120kW (160hp) Mercedes D.III 6-cylinder water-cooled inline piston engine
Speed: 160 km/h (99 mph)
Endurance: 3 hours
Ceiling: 2890m (9480ft)
Crew: 1
Armament: Two 7.92mm (0.312in) LMG 08/15 'Spandau' machine guns

---

Albatros W.4

W.4 number 965 was sent to the Baltic in February 1917 to operate from Windau Air Station. By this time the Navy had realised that two-seater floatplanes offered greater versatility and the W.4 would eventually be supplanted by the superlative Hansa-Brandenburg seaplanes.

Hansa-Brandenburg W.12

A versatile and formidable aircraft, the W.12 proved to be a particularly successful maritime fighter in long-range skirmishes with RNAS flying boats.

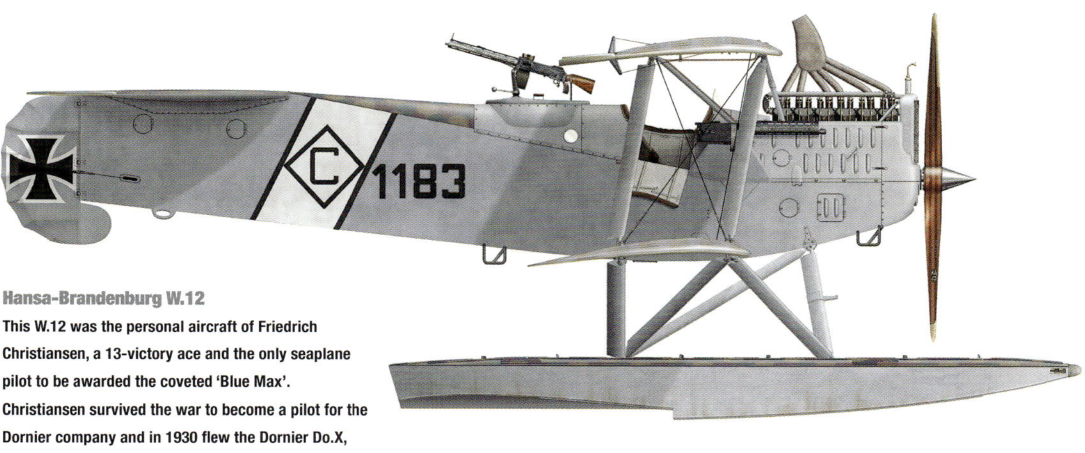

Hansa-Brandenburg W.12

This W.12 was the personal aircraft of Friedrich Christiansen, a 13-victory ace and the only seaplane pilot to be awarded the coveted 'Blue Max'. Christiansen survived the war to become a pilot for the Dornier company and in 1930 flew the Dornier Do.X, then the world's largest aircraft, on its maiden flight to New York.

Representing a significant leap in capability over single-seat floatplane fighters such as the Albatros W.4, the Hansa-Brandenburg W.12 was designed by Ernst Heinkel during late 1916 and made its first flight in February the following year. Featuring a distinctive deep, slab-sided, upswept fuselage that compensated for the side area of the floats, the most striking aspect of the design was the rudder that protruded below the aircraft, rather than above as is more conventional. The reason for this was to give the rear gunner an unparalleled field of fire, a feature improved still further by the cantilever horizontal tail surfaces that required no struts or bracing. The wings also featured no wire bracing and the gunner could fire between, or even through, them without fear of breaking a bracing cable.

Fitted with either a Benz Bz.III or Mercedes D.III engine (Mercedes-powered machines featured a radiator in the leading edge of the top wing, whereas aircraft with Benz engines had a nose radiator), the W.12 possessed good performance and excellent manoeuvrability although various problems delayed the acceptance of the type until August. By this time the first production aircraft, which had been ordered before the prototype had even flown, had started to appear and the aircraft entered service in the same month. Criticism of the handling of early aircraft led to the adoption of a lengthened fuselage on later production aircraft and ailerons on the lower wing in addition to those on the top plane. There were also fears that the cockpit's placement, under the wing and surrounded by struts, would cause difficulty for the pilot in

Hansa-Brandenburg W.12 (late production)

Weight: (gross) 1454kg (320lb)
Dimensions: Length 9.6m (31ft 6in) Wingspan 11.2m (36ft 9in) Height 3.3m (10ft 10in)
Powerplant: One 110kW (150hp) Benz Bz.III 6-cylinder water-cooled inline piston engine or one 120kW (160hp) Mercedes D.III 6-cylinder water-cooled inline piston engine
Speed: 161km/h (100mph)
Endurance: 3.5 hours
Ceiling: 5000m (16,400ft)
Crew: 2
Armament: Two 7.92mm (0.312in) LMG 08/15 'Spandau' machine guns fixed firing forward, one 7.92 mm Parabellum MG14 machine gun flexibly mounted in rear cockpit

Hansa-Brandenburg W.12

Crewed by Lt Urban and Lt Ehrhardt, this W.12 was one of a group of three aircraft that located and destroyed the British non-rigid airship C27 on 11 December 1917. Credit for the victory is attributed to Christiansen but Ehrhardt took a series of photographs of the action from this machine, one of very few genuine examples of air combat photography during World War I.

extricating himself in the event of an emergency, and later aircraft featured an enlarged cut-out above the pilot's seat to allow a more rapid egress, a future that also improved pilot view.

Communications upgrade

Operations proved that the new aircraft was a great success. The two-man crew allowed for a much more useful patrol aircraft and later W.12s were fitted with radio, allowing for great operational flexibility. For example, W.12s could halt a ship attempting to run the naval blockade by dropping bombs in its path or firing across its bow and then radio for a prize crew to intercept the vessel and take it to port. However, the aircraft was well able to deal with any enemy it was likely to meet in aerial combat. The loss of British airship C27 to W.12s soon after the latter's introduction caused the RNAS to restrict airship operations

to areas where German aircraft were unlikely to appear. Furthermore the powerful British flying boats that had hitherto patrolled alone with relative impunity over the North Sea were now compelled to operate in flights of three to make sure that at least one of them would complete the mission and return to base. Given the effect on its enemies, it is unsurprising therefore that the W.12 was popular with its crews, who nicknamed it the 'Kamel', possibly due to the oddly shaped fuselage.

Despite the appearance in service of the monoplane W.29, the W.12s served until the armistice in German service. However, the end of the war did not end their career, nor even production of the W.12, as a further 35 aircraft were built under licence in the Netherlands as the Van Berkel W-A. These would remain in service with the Dutch Naval Air Service until 1933.

Hansa-Brandenburg W.19

Identical to the W.12 in general configuration, the W.19 was a considerably larger aircraft designed to fulfil a requirement for an aircraft in the same class but with greater range.

Fitted with the Maybach Mb.IVa, the W.19 was considerably more powerful than the W.12, enabling it to possess a similar performance despite its greater size and weight. Believed to have been flown for the first time in August 1917, the W.19 started to be distributed to front-line units in the spring of 1918, with the first operational use of the new aircraft occurring in May.

Service pilots considered the W.19 to be easier to fly than the W.12, but conceded that it lacked the manoeuvrability of the earlier machine. Although it was intended to replace the W.12 with the W.19, in reality the new aircraft generally supplemented rather than replaced the W.12 on operations, although at least one unit was completely equipped with the W.19.

Impressive endurance

With its impressive endurance, the W.19 was able to scout for targets and then radio details to shorter-ranged W.12 and W.29s waiting on the surface. However, the W.19 was also a perfectly capable combat machine in its own right: a particularly effective action on 11 August saw a force of W.19s, in concert with W.12s and W.29s, attack a group of six British motor torpedo boats (MTBs), sinking three and forcing the others to beach in Holland, where the crews were interned.

The W.19 also fared well in combat with the well-armed RNAS Felixstowe flying boats operating in the North Sea. It had been intended to supplement and replace the W.19 with the monoplane W.33, but only six of these had been manufactured by the end of hostilities. Subsequently at least two W.19s saw postwar service with Italy and one was delivered to Japan.

Hansa-Brandenburg W.19

Weight: (gross) 1454kg (3206lb)
Dimensions: Length 10.6m (34ft 11in) Wingspan 13.8m (45ft 3in) Height 4.1m (13ft 5in) Powerplant: One 190kW (250hp) Maybach Mb.IVa 6-cylinder water-cooled inline piston engine
Speed: 151km/h (94mph)
Endurance: 5 hours
Ceiling: 5000m (16,400ft)
Crew: 2
Armament: Two 7.92mm (0.312in) LMG 08/15 'Spandau' machine guns fixed firing forward, one 7.92mm Parabellum MG14 machine gun flexibly mounted in rear cockpit

---

Hansa-Brandenburg W.19

The W.19 was structurally similar to the W.12, although it did require wire bracing on the outboard sections of its wings, leading to a curious 'half-finished' look, the inner bays being devoid of bracing wires, like the W.12. Aircraft 2207 was the first production example of the type.

Hansa-Brandenburg W.29

Seeking to improve on the W.12, Hansa-Brandenburg produced the monoplane W.29. An outstanding aircraft, its service life in German hands was brief although it enjoyed a lengthy post-war service with other nations.

Utilising a semi-cantilever monoplane wing of roughly the same area as the combined two wings of the W.12 from which it was derived to minimise drag, the W.29 achieved a useful increase in performance despite using the same engine and carrying the same payload. The fuselage remained largely unaltered and with the removal of the upper wing the field of fire of the observer, already excellent on the W.12, was now even better. The aircraft was tested during the spring of 1918, entering production in April. Production aircraft used the Benz Bz.III engine, probably because supplies of the slightly more powerful Mercedes D.III were earmarked for higher priority landplane requirements.

Later production aircraft received the considerably more powerful Benz Bz.IIIa and eventually 11 W.29s were built with the superlative BMW IIIa. The only problematic area of the W.29, like the W.12 before it, was its wooden floats that were constantly prone to leaks. Despite being painted with a waterproof bituminous paint, the floats had to be changed regularly, although the W.29s received superior duralumin floats towards the end of the conflict.

Submarine strike

Given that the W.29 took the winning formula of the W.12 and improved upon it, it was no surprise that the new seaplane proved immediately effective. For example, the first five examples issued to Seeflugstation

---

Hansa-Brandenburg W.29

Weight: (gross) 1494kg (3294lb)

Dimensions: Length 9.36m (30ft 9in) Wingspan 13.5m (44ft 3in) Height 3m (9ft 10in)

Powerplant: One 110kW (150hp) Benz Bz.III 6-cylinder water-cooled inline piston engine

Speed: 175km/h (109mph)

Endurance: 4 hours

Ceiling: 5000m (16,400ft)

Crew: 2

Armament: Two 7.92mm (0.312in) LMG 08/15 'Spandau' machine guns fixed firing forward (one if radio equipment fitted), one 7.92mm Parabellum MG14 machine gun flexibly mounted in rear cockpit; up to four 5kg (11lb) bombload could be carried in rear cockpit

---

Bearing the name 'Anne', this W.29 was stationed at Zeebrugge as part of Seeflugstation Flandern II in early April 1918. The W.29 did away with the large exhaust manifold of the W.12 and W.19 and instead featured six stub exhausts discharging directly from each cylinder.

Hansa-Brandenburg W.29

The diagonal fuselage stripes denote that this W.29 was on the strength of the Seeflugstation on the resort island of Nordeney. This W.29 features the thinner *balkankreuz* markings that were introduced from June 1918; the different shade where the earlier markings have been painted over is apparent on the fuselage.

Flandern I to reach active service were ferried by Friedrich Christiansen and other crews to their base at Zeebrugge on 1 July 1918, and the W29s were flown on operations the following morning. Within a week of the W.29's appearance, Christiansen's W.29s had surprised the British submarine C25 on the surface near Harwich and attacked it. Their machine gun fire killed the captain and pierced the pressure hull, crippling the unfortunate vessel, and seaplanes maintained attacks on a second submarine, E51, and the destroyer HMS *Lurcher* as they attempted to take the C25 – 'leaking like a sieve' – in tow. The C25 survived (although the German airmen believed they'd sunk it), but this attack demonstrated the W.29s' ability to interfere with Royal Navy operations, even in home waters. For the remaining four months of the war the W.29s effectively supplemented the W.12s with naval units, engaging in maritime patrol and tussling with the Felixstowe flying boats of the RNAS.

Built under licence

In total, 239 aircraft were ordered for the Kaiserliche Marine, of which 209 were built before the armistice – although plans were also afoot at the end of the war to build the W.29 in Austria-Hungary under licence by UFAG with the excellent Porsche-designed 138kW (185hp) Austro-Daimler 6. Only one was built before the end of the war but a further two were completed in 1919 by the short-lived Hungarian Soviet Republic and served with the 9th floatplane squadron at Cspel near Budapest.

Japanese service

Post-war, 15 W.29s also saw extensive service with Japan, which received the aircraft as war reparations. The W.29 was then put into large-scale production by Japan, some 64 being built by Nakajima and 92 by Aichi as the Hansa-shiki Suij Teisatsu-ki (Hansa Model Reconnaissance Seaplane).

Despite relative unpopularity amongst Japanese crews, these served into the 1930s, latterly as trainers, and several then joined the civil register, some modified with a cabin for four passengers.

Similarly, after purchasing one example from Germany in 1919, 15 W.29 copies were produced in Denmark at the Orlogsværftet (Danish Royal Naval Dockyard) as the HM.1 between 1921 and 1927, remaining in service until 1930.

Zeppelin airships

An iconic weapon of World War I, the Zeppelin dirigible airships were emblematic of a new and terrifying era of warfare: the strategic bombing of civilian targets. Physically huge and seemingly inexorable, these enormous aircraft inspired a reaction out of all proportion to the material damage they caused.

Count Ferdinand von Zeppelin's interest in aviation was seemingly sparked by an ascent he made in an observation balloon while attached to the Union Army as an official observer during the American Civil War. His first rigid airship flew in 1900 and by the outbreak of war Luftschiffbau Zeppelin GmbH had manufactured 25 large airships, several being used very successfully as passenger aircraft by the world's first airline, DELAG.

As a military machine the rigid airship was something of a mixed bag: for example, from 1908 onwards Army airship Z I had made several long-duration flights; conversely, in 1913, Navy airship L 2 had leaked some of its own hydrogen into one of its engines and caught fire, killing all 28 aboard.

Nonetheless, the rigid airship was capable of sustained flight duration while carrying payloads that could only be dreamed of by heavier than air aircraft designs. For example, LZ 4 completed a 12 hour flight as early as July 1908 when the world record for aeroplanes was a mere 59 minutes.

By the outbreak of war Zeppelin had been joined by the Schütte-Lanz company as a manufacturer of large airships. In some respects the Schütte-Lanz ships were more advanced than their Zeppelin counterparts, with a more aerodynamic overall form imparting greater speed.

Schütte-Lanz also introduced the use of efficient cruciform tail surfaces (later adopted by Zeppelin) but its airships were generally less well thought of. Zeppelin airships utilised an internal framework made of duralumin but Schütte-Lanz favoured a glued wood and plywood construction, meaning they were more prone to the effects of moisture, a particularly problematic feature for Naval airships. Despite this, the Schütte-Lanz airships were used operationally by both services.

Assault on London

Army Zeppelin Z VI carried out the first airship bombing mission of World War I when it attacked Liège on 6 August 1914, using artillery shells as no bombs were available. In the first few months of 1914 the Zeppelin and

R-class Zeppelin

Weight: (gross) 63,907kg (140,891lb)
Dimensions: Length 198m (649ft 4in) Diameter 23.9m (78ft 4in)
Powerplant: Six 183kW (245hp) Maybach HS Lu high compression 6-cylinder water-cooled inline piston engines
Speed: 103km/h (62mph)
Range: 7400km (4600 miles)
Ceiling: 3900m (12,500ft)
Crew: 10
Armament: Three 7.92mm (0.312in) Maschinengewehr 08 (Navy) or 7.92mm (0.312in) Parabellum MG14 (Army) machine guns flexibly mounted in forward dorsal position atop hull, one in rear dorsal position, two in each front and rear gondolas and one in each of the ventral engine nacelles; up to 7600kg (16,755lb) bombload

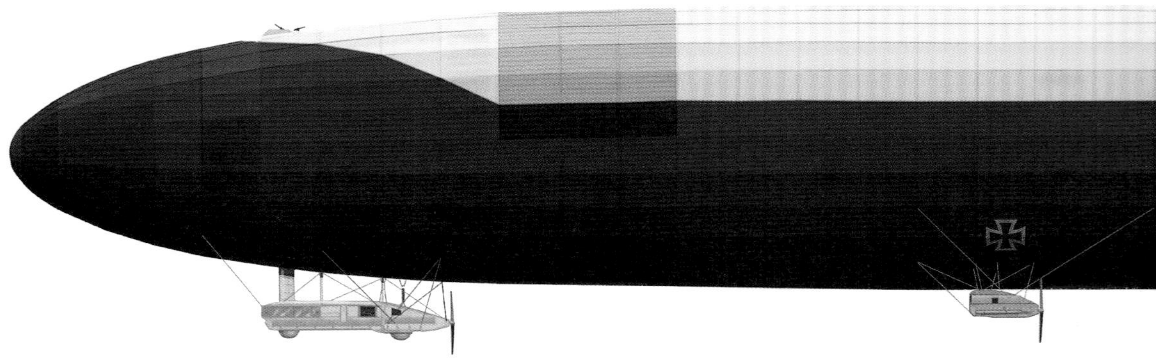

Schütte-Lanz airships were kept busy in tactical reconnaissance and bombing missions on both Western and Eastern Fronts. Antwerp was bombed on 25 August, Warsaw on 24 September and Paris was attacked for the first time 21 March 1915. London's turn came on 8 September 1915. By this time the British Isles had already been bombed by airships – in January, Great Yarmouth and Sheringham on the Norfolk coast were bombed after the airship crews failed to find the Humber dockyards, their intended targets, some 180km (112 miles) to the north, inflicting relatively little damage. The following month Kaiser Wilhelm II approved attacks on London that he had hitherto forbidden due to fears that his relatives in the British royal family might be killed or injured, although bombing was supposed to be restricted to an area east of Charing Cross.

London raids occurred fairly regularly through the spring of 1916 until Navy Zeppelin L13, captained by Heinrich Mathy, flew her dramatic 8 September raid, dropping bombs on a variety of locations as she appeared to nonchalantly cruise over the capital. "It gave an impression of absolute calm and absence of hurry," remarked one witness. In the course of a few minutes, L 13's bombs had caused more than £500,000 of damage and killed 22 people. It was material damage that would never be equalled in any other bombing mission, and this single raid resulted in over a sixth of the total damage inflicted by airships during the entire war. In addition, one British aircraft was lost attempting to intercept the Zeppelin. Other raids followed leading to a peak of airship activity over London during 1916.

Damaged morale

It is hard to exaggerate the effect the German airships had on British national morale. The raids provoked a feverish effort to improve London's air defences. As early as 7 June 1915, Reginald Warneford had managed to destroy an airship over Belgium by dropping bombs on it, but on 3 September 1916 William Leefe Robinson shot down Schütte-Lanz SL 11 over British soil and for the rest of the war the giant ships proved increasingly vulnerable. Army airship operations were wound down after the loss of SL 11, but the Navy pressed on and suffered ever greater losses. A total of 10 airships were taken down during 1916, including that of Heinrich Mathy who was killed with his crew when L 31 was shot down by a Sopwith Camel in the early hours of 2 October 1916. The Germans responded with the 'Height Climbers' – lightened Zeppelins designed to operate above the ceiling of any defending fighters. Unfortunately for their crews the latest British aircraft were perfectly capable of reaching their operating altitude, and on 5 August 1918 the final airship raid of the war saw the latest Zeppelin L 70 shot down by a DH.4 crewed by Egbert Cadbury and Robert Leckie, both of whom had already destroyed one airship each. None of the crew survived and of the other three Zeppelins attempting to attack that night, none dropped a single bomb on British soil. The head of the Navy's airship service Peter Strasser was on board L 70 on its final mission and his loss effectively ended the strategic bombing campaign by airship.

By November 1918 Zeppelin had supplied over 100 rigid airships to the military with Schütte-Lanz adding around 20 more.

R Class Zeppelin L 33

Zeppelin L 33 was one of the R class, or 'Super Zeppelins', which first appeared in mid 1916 and were notably more streamlined than their predecessors. L 33 however was holed by anti-aircraft fire on its maiden mission and crashed-landed in Essex where the crew were taken prisoner.

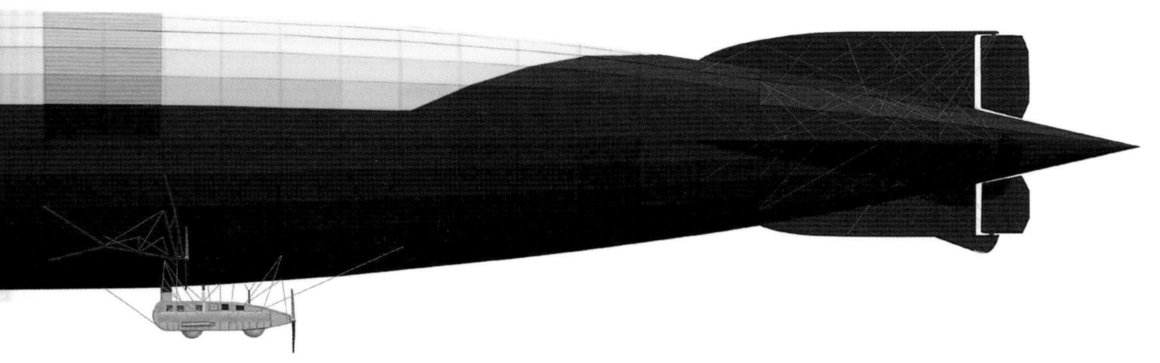

IdFlieg designation system

A – Unarmed reconnaissance monoplane

B – Unarmed reconnaissance biplane

C – Armed two-seat biplane

CL – Light two-seater, initially intended as escort fighters but later utilised mainly for ground attack

D – Single-seat, armed biplane, but later any fighter

Dr – Dreidecker – triplane fighter (initially F)

E – Eindecker – monoplane fighter, this designation initially included armed monoplane two-seaters. New monoplane types at the end of the war designated as D (single seat) or CL (two seat).

G – Grossflugzeug – Large twin-engined types (originally K for Kampfflugzeug)

GL – Lighter twin-engined bombers

J – Schlachten – Fuel tanks, pilot and engine protected by armour plate

N – C-type aircraft adapted for night bombing

R – Riesenflugzeug – 'Giant' aircraft – at least three engines, usually more, all serviceable in flight

Glossary

Bildaufklärer:
Photo-reconnaissance aircraft

Bogohl:
Bombengeschwader der Oberste Heeresleitung: Bomber Wings under direct control by the German Army's High Command

Eindecker:
Monoplane

Doppeldecker:
Biplane

Dreidecker:
Triplane

Feldflieger Abteilung (FFA):
Field Flier Detachment, the initial units of the German Army in 1914–1915

Flieger-Abteilung (Artillerie) (FA(A)):
Flier Detachment (Artillery)

Fliegertruppe:
Die Fliegertruppen des deutschen Kaiserreiches (Imperial German Flying Corps), the name of the air arm of the Imperial German Army until October 1916

Fokkerstaffel:
Early fighter units predating the formation of the Jagdstaffeln equipped with Fokker Eindeckers, hence the name

Grossflugzeug:
'Large aircraft', usually intended for bombing

IdFlieg:
Inspektion der Fliegertruppen (Inspectorate of Flying Troops) was the bureau of the German Army that oversaw military aviation

Jagdgeschwader:
'Hunting Wing', a wing of fighter units

Jagdstaffel (Jasta):
'Hunting Squadron,' a fighter unit

Kagohl:
Kampfgeschwader der Oberste Heeresleitung: Tactical Bomber Wings under direct control by the German Army

Kaiserliche Marine:
Imperial German Navy

Kampfflugzeug:
'Fighting Aircraft', an early concept for a large aircraft intended for aerial combat

KEK:
Kampfeinsitzerkommando, Combat Single-Seater Command, a predecessor unit to the Jagstaffeln

Kest:
Kampfeinsitzer Staffel, Combat Single-Seater Squadron, another predecessor unit to the Jagdstaffeln

Kogenluft:
Kommandierender General der Luftstreitkräfte, Commanding General of the Air Service

Luftskreitkräfte:
Imperial German Air Service and all associated aviation activities such as anti-aircraft, home defence and air intelligence were unified as the Luftstreitkräfte in October 1916

Marine-Landfliegerabteilung:
Naval Land Flier Detachment, a Naval unit operating on land alongside regular army units

Reisenflugzeug:
'Giant aircraft,' specifically an aircraft with at least three engines, all serviceable in flight

Riesenflugzeugabteilung:
Giant Aircraft Detachment

Schlachtstaffel (Schlasta):
'Battle squadron', a ground attack unit

Schutzstaffel (Schusta):
'Protection squadron', an escort fighter unit

Seeflugstation:
Naval Air Station

Index

Picture Credits

Photographs:

AirSeaLandPhotos: 6, 7, 35, 53, 59, 60, 70, 72, 82, 86, 89, 93, 98, 102, 108, 109, 112, 113

Getty Images: 8 (Keystone-France/Gamma-Rapho)

Artworks:

All artworks copyright **Ronny Bar,** except for the following:

Amber Books: 50/51, 67, 101 top, 112

Edward Ward: 120/121